FOOD & DRINK –
GOOD MANUFACTURING PRACTICE

A GUIDE TO ITS RESPONSIBLE MANAGEMENT

FOOD & DRINK –
GOOD MANUFACTURING PRACTICE
A GUIDE TO ITS RESPONSIBLE MANAGEMENT

Seventh Edition

Institute of Food Science and Technology (UK)
5 Cambridge Court
210 Shepherd's Bush Road
London W6 7NJ

Registered Offices
John Wiley & Sons, Inc., 111 River Street, Hoboken, NJ 07030, USA
John Wiley & Sons Ltd, The Atrium, Southern Gate, Chichester, West Sussex, PO19 8SQ, UK

Editorial Office
9600 Garsington Road, Oxford, OX4 2DQ, UK

For details of our global editorial offices, customer services, and more information about Wiley products visit us at www.wiley.com.

Wiley also publishes its books in a variety of electronic formats and by print-on-demand. Some content that appears in standard print versions of this book may not be available in other formats.

Library of Congress Cataloging-in-Publication Data

Names: The Institute of Food Science & Technology Trust Fund, author.
Title: Food & drink–good manufacturing practice : a guide to its responsible management /
    The Institute of Food Science & Technology Trust Fund [author].
Other titles: Food & drink
Description: 7th edition. | Hoboken, NJ, USA : Wiley, [2018] | Revised edition of Food & drink. |
    Includes bibliographical references and index. |
Identifiers: LCCN 2018016214 (print) | LCCN 2018016958 (ebook) | ISBN 9781119388531 (Adobe PDF) |
    ISBN 9781119388524 (ePub) | ISBN 9781119388449 (pbk.)
Subjects: LCSH: Food industry and trade–Great Britain–Quality control.
Classification: LCC TP369.G7 (ebook) | LCC TP369.G7 F66 2018 (print) | DDC 664.00941–dc23
LC record available at https://lccn.loc.gov/2018016214

Cover Design: Wiley
Cover Image: © Mark Yull/Shutterstock

Set in 9/10.5pt TimesNewRoman by SPi Global, Pondicherry, India
Printed in Singapore by C.O.S. Printers Pte Ltd

10 9 8 7 6 5 4 3 2 1

Foreword

In the United Kingdom we all depend upon the food industry, our largest manufacturing sector, to deliver high standards when it comes to food that is safe to eat, and food that is what it says it is. We all want the public to be able to trust the food they eat, at home and whilst out and about. It is the industry's responsibility to ensure that food is safe and authentic, and to deliver and sustain strong foundations for public trust in food.

The 7th edition of the IFST Guide to Good Manufacturing Practice is an important resource for food businesses. With thorough and detailed guidance, it sets out how businesses meet their legal obligations and shows how to deliver public confidence and trust. I welcome the increased focus on food authenticity and integrity in this edition, which will help to protect businesses and consumers alike from the risks of food fraud. It is an area we at the Food Standards Agency are also focused on. We wish to extend the remit and scale of the National Food Crime Unit, working in partnership with other agencies, and we are revitalising our surveillance approach to keep pace with constant change in our national and global food system. Regulation needs to adopt new and innovative ways of delivering, and to keep refreshing the relationships between regulator and industry to succeed against new threats and challenges. This guide will help business get it right, and help us all deliver a safe, secure and trusted food sector into the future.

Heather Hancock
Chair of the Food Standards Agency

CONTENTS

Preface to the Seventh Edition   ix
Acknowledgements   xi
Decision Makers' Summary   xiii

Part   **I**   **General Guidance**
Chapter   1   Introduction   3
               2   Quality Management System   9
               3   Hazard Analysis Critical Control Point   17
               4   Food Safety Culture   31
               5   Food Crime and Food Integrity Management Systems   35
               6   Food Crime Risk Assessment   45
               7   Security and Countermeasures   53
               8   Food Toxins, Allergens and Risk Assessment   63
               9   Foreign Body Controls   75
               10   Manufacturing Activities   85
               11   Management Review, Internal Audit and Verification   95
               12   Product and Process Development and Validation   101
               13   Documentation   107
               14   Product Identification and Traceability   115
               15   Provenance and Authenticity   121
               16   Electronic Identification and Digital Traceability Techniques   125
               17   Personnel, Responsibilities and Training   129
               18   Worker Welfare Standards   147
               19   Premises and Equipment   151
               20   Water Supply   169
               21   Cleaning and Sanitation   173
               22   Infestation Control   183
               23   Purchasing   191
               24   Packaging Materials   197
               25   Smart Packaging   201
               26   Internal Storage   205
               27   Crisis Management, Complaints and Product Recall   211
               28   Corrective and Preventive Action   217
               29   Reworking Product   221
               30   Waste Management   225
               31   Food Donation Controls and Animal Food Supply   229
               32   Warehousing, Transport and Distribution   233
               33   Contract Manufacture and Outsourced Processing and Packaging   239
               34   Calibration   241
               35   Product Control, Testing and Inspection   245
               36   Provenance and Integrity Testing   251
               37   Labelling   253

	38	Good Control Laboratory Practice and Use of Outside Laboratory Services	257
	39	Electronic Data Processing and Control Systems	265
	40	Sustainability Issues	269
	41	Environmental Issues	271
	42	Health and Safety Issues	275

Part | **II** | **Supplementary Guidance on Some Specific Production Categories** |
Chapter | 43 | Heat-Preserved Foods | 279 |
	44	Chilled Foods	289
	45	Frozen Foods	299
	46	Dry Food Products and Materials	307
	47	Compositionally Preserved Foods	311
	48	Foods Critically Dependent on Specific Ingredients	313
	49	Irradiated Foods	315
	50	Novel Foods and Processes	321
	51	Foods for Catering and Vending Operations	325
	52	The Use of Food Additives and Processing Aids	327
	53	Responsibilities of Importers	331
	54	Export	333

Part | **III** | **Mechanisms for Review of this Guide** |

Appendix I | Definition of Some Terms Used in this Guide | 337 |

Appendix II | Abbreviations Used in this Guide | 347 |

Appendix III | Legislation and Guidance | 355 |

Appendix IV | Additional References | 357 |

Appendix V | Contribution to the Seventh and Previous Editions of the Guide | 361 |

Preface to the Seventh Edition

The 6th edition built on previous editions and provided additional content in the area of food safety management systems and quality management systems, their design, validation, implementation and verification. Consideration of what good manufacturing practice (GMP) is, and evolves to be, has led to the need for a number of new chapters in this 7th edition. The melamine incident in China, fipronil in eggs and the horsemeat incident 2013 in Europe has caused food manufacturing organisations, the food supply chain and those involved in wider food policy development and implementation to consider issues around food integrity, food crime and general malpractice in the food chain and how protocols to mitigate risk need to be included within GMP. The rapid development in testing programmes to demonstrate the provenance of food materials and food products means that manufacturers can now more easily verify labelling information and the claims they make on their products. As a result, the 7th edition focuses on the growing interest in food integrity management systems and how manufacturers need to demonstrate they have done everything reasonable to ensure the integrity of their products, the processes they employ, the data and information they often rely upon, and the people who undertake the tasks critical to ensuring food safety, legality and quality.

Louise Manning
7th Edition

Acknowledgements

A list of many of the organisations and individuals from whom help, information or comment has been received for this edition is presented as Appendix V. This is inevitably incomplete and cannot include acknowledgement of numerous verbal comments received. However, I welcome the opportunity to thank all who participated and particularly the members, both past and present, of the GMP Working Groups. Especially, I would thank Professor J.R. Blanchfield, who as Editor and Convener of the GMP Working Group, 4th edition, made an enormous contribution to the development of the 5th and 6th editions of this Guide.

As with the previous editions that I have edited, the preparation of this 7th edition has been an enjoyable and thought-provoking experience.

Louise Manning
7th Edition

Decision Makers' Summary

This summary is especially addressed to the decision makers within food and drink company chairmen, presidents, chief executives, directors and general managers, who are not normally directly involved in detailed design and implementation of good manufacturing practice (GMP) systems, but whose responsibility it is to establish GMP policies and strategies for their companies, and to provide the necessary authority, facilities and resources to the functional managers and staff to implement the requirements effectively.

In this Guide, GMP is considered as that part of a food and drink control operation that is aimed at ensuring that products are safe, legal, meet integrity criteria and are of the required quality. Effective GMP ensures that products are consistently manufactured to a quality appropriate to their intended use. It is thus concerned with manufacturing practices, food safety, legality, quality and integrity management systems.

The ever-increasing interest among consumers, retailers, enforcement authorities and other stakeholders such as insurers or shareholders in the conditions and practices employed in food manufacture and distribution heightens the need for the food manufacturer to operate with a clearly defined strategy with high-level policies together with associated operational procedures and protocols. The ability to demonstrate that the principles and measures identified in this Guide have been fully and effectively implemented could, in the event of a product incident, consumer complaint or formal prosecution, assist the manufacturer in demonstrating that all reasonable steps had been taken to prevent the cause of the incident from occurring, or indeed avoid an offence being committed. Enlightened self-interest alone should persuade food manufacturers to follow these guidelines.

The manufacturer of a food product must comply with the relevant legal requirements of the country for which the food is intended, for example those of composition, safety, hygiene and labelling. While fulfilling these, however, she/he has a concept of the market at which she/he is aiming and its requirements (e.g. in the case of a food or drink product, its appearance, flavour, texture, presence or absence or amount of particular nutritional components, inbuilt convenience, shelf life, presentation and price). These factors determine the formulation, processing and packaging of the product. Product quality is defined in a product specification that should encompass all of these requirements and express them in a clear, unambiguous manner.

The retailer may approach a manufacturer with a new product concept and request that a manufacturer design a product or process to meet the specific criteria. Of course, the manufacturer's assessment of what the market wants may be correct or incorrect. While the concept effectively meets all of the law's requirements, it may, or may not, effectively meet purchasers' expectations, but unless and until the manufacturer or retailer changes it, the product specification remains the standard with which the product should conform, and GMP is designed to achieve this.

Uniform conformance with product specifications is difficult with food and drink products. The main raw materials for food and drink manufacture derive from nature and are subject to natural variations. In primary production, wide variations may occur among cultivars and also because of seasonal, weather and cultivation differences. In animals, apart from differences between individuals, variance between breeds and rearing systems leads to the potential for inconsistency. Therefore the additional task of the food or drink manufacturer, aided by the knowledge and skills of food science and technology, is to make a reasonably uniform product from variable raw materials by an appropriate combination of raw material selection, raw material pretreatment, formulation adjustment and processing out variation which is outside the boundaries of the product specification.

The Basis for GMP

GMP has two complementary and interacting components: the manufacturing operations and wider management systems [which, for the purposes of this Guide, the Institute of Food Science & Technology (IFST) has designated 'food control'] (see Figure 1). Both these components must be well designed and effectively implemented. The same complementary nature and interaction must apply to the respective management of these two functions, with the authority and responsibilities of each clearly defined, agreed and mutually recognised. This is not to disregard the importance of other key functions essential to the effective functioning of a company, or indeed of those functions contributing direct services or advice to the manufacturing operation (e.g. purchasing, cost accounting, work study, production planning and engineering maintenance). These terms are explored in more detail in Chapter 2.

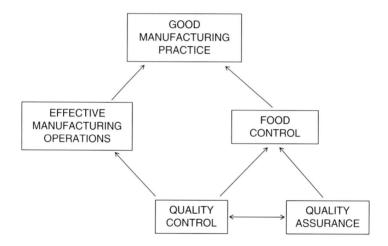

Figure 1

What constitutes 'well designed' in these two contexts mentioned above is not just a matter of common sense, or something that would be self-evident to non-technical business people. As well as management skills, it also involves extensive and up-to-date knowledge of current and emerging quality issues, food safety hazards and best practice in terms of food science and technology relating to the ingredients, processes, packaging and products concerned.

Effective Manufacturing Operations

GMP requires that every aspect of manufacture is fully defined in advance and that all the resources and facilities are specified, namely:

- specific measures undertaken at critical control points (CCPs) based on food safety hazard analysis and food integrity threat analysis, or critical quality points (CQPs) identified in the quality planning process;

- adequate design of premises and suitable manufacturing and storage space;
- suitable process flow with process design to streamline the process and minimise the potential for cross-contamination;
- correct and adequately maintained equipment;
- appropriately trained people;
- correct raw materials, processing aids and packaging materials;
- appropriate storage and transport facilities;
- documented operational procedures and cleaning schedules;
- appropriate management and supervision; and
- adequate technical, administrative and maintenance services

are provided, in the right quantities, at the right times and places, and are utilised as intended. In order to ensure that operations do proceed according to plan, it is also necessary to:

- provide operators with documented procedures in clear unambiguous instructional language (with due regard to reading, numeracy and language problems);
- train and motivate the operators to carry out the procedures correctly;
- undertake formal review to ensure that training and instruction have been effective;
- avoid, if possible, incentive bonus schemes, but, if unavoidable, to build into any incentive bonus schemes adequate safeguards against unauthorised 'short cuts' or trade-offs;
- provide a food control programme working along the lines indicated below;
- ensure that genuine records are completed during production and that they demonstrate that specified procedures were in fact complied with, and to enable the history of manufacture and distribution of a batch subsequently to be traced should a problem arise or a product withdrawal or recall be necessary;
- establish a well-planned and effective system to carry out a product withdrawal or recall, should that prove necessary; and
- establish a tried and proved business continuity and crisis management procedure in case of need.

Effective Food Control

The other and complementary major component of GMP is effective food control. Effectiveness requires:

- well-qualified and appropriately experienced individuals working in food control management participating in the development and validation of process controls and specifications that address the safety, legality, quality and integrity of food;
- competent staff and adequate facilities to undertake all the relevant inspection, sampling and testing of materials, and monitoring of process conditions and relevant aspects of the production environment (including all aspects of hygiene) and management of potential food safety hazards and food integrity threats;
- verification activities that are developed and implemented by appropriately experienced personnel in order to demonstrate that the food products and the process are consistently under the appropriate level of control and identify areas for preventive action where system weaknesses and vulnerabilities are detected; and
- rapid feedback of information (accompanied where necessary by advice) to manufacturing personnel, thereby enabling prompt adjustment or corrective action to be taken and enabling processed material to be approved as fit for either further processing or sale, or to be segregated for decision as to appropriate disposition, for example reject, regrade or reprocessing.

Responsible Management

Of course, the requirements of effective manufacturing operations and of effective food control mentioned above are merely headings and within each there are very many aspects that are considered more fully within the body of this Guide. The Institute hopes that the Guide will prove of help to the management of food and drink companies, to those concerned with private and public verification activities, food law enforcement and consumer protection, to the students who will be the food technologists, engineers and production managers of tomorrow and to those responsible for training them.

The full title of the Guide is *Food & Drink – Good Manufacturing Practice: A Guide to its Responsible Management*. The reference to responsible management is deliberate. GMP can only stem from policy firmly and uncompromisingly stated and continuously pursued by a company board and general management, which, moreover, provides adequate physical, financial and human resources for the purpose.

PART I – GENERAL GUIDANCE

1   INTRODUCTION

1.1   The purpose of this Guide is to outline the responsibilities of managers in relation to the efficient manufacture and control of food and drink products, thereby ensuring that such products are safe, wholesome and of the nature and quality intended. While it addresses manufacture of food and drink for use in the retail, catering and vending industries, it does not deal with catering and retail activities per se. The Guide is therefore particularly concerned with management practices associated with:

- factors affecting product safety, product legality, product integrity and product quality;
- product manufacture in terms of product and process control and handling of food under hygienic conditions in conformity with product, packaging and labelling specifications; and
- matters such as training of personnel, documentation and record keeping, supplier approval, suitability of premises and equipment and site standards, waste avoidance, recovery and reworking of materials, laboratory management, traceability, verification activities, and preventive and corrective action and the management of customer complaints and product recall.

1.2   It is emphasised that the Guide is concerned with advice based on principles of good manufacturing practice (GMP), and it is recognised that methods other than those described, but which achieve the same ends, may be equally acceptable. Personnel and premises hygiene, because of its importance, is treated as a continuous theme and a subject for consideration throughout the document.

The Guide is in three parts:

Part I: deals with matters of general application;
Part II: deals with guidance on specific manufacturing and/or food categories; and
Part III: covers mechanisms for review of the Guide.

1.3   The Guide does not deal directly with such matters as operative safety and welfare, ethical matters, animal welfare or environmental issues including water and energy conservation. It refers to resource management and waste control, engineering, maintenance and transport and distribution only in respect of those aspects that have a bearing on manufacturing practices. In general it does not deal with matters unrelated to scientific,

Food & Drink – Good Manufacturing Practice: A Guide to its Responsible Management, Seventh Edition.
The Institute of Food Science & Technology Trust Fund.
© 2018 John Wiley & Sons Ltd. Published 2018 by John Wiley & Sons Ltd.

technological and organisational aspects affecting product safety, product legality, product integrity and product quality.

1.4   The Guide has been written at a time when the United Kingdom (UK) is negotiating a new relationship with the European Union (EU) in terms of legislative harmonisation and trade agreements. To this end legislation still in force at the time of writing has been referenced, but may change post publication of this Guide. This is true of any legislation and/or policy approach in a given country or trading group when referenced by a static publication. The principles of GMP are universal and in many ways transcend the specifics of one nation's current or emerging legislation. Food manufacturers supplying internationally need to be aware of not only the legislation in the country in which they are manufacturing, but also the need for the products produced to comply with the legislation in countries to which they seek to export. There are many instances of product recalls in countries as a result of food products, for example, not complying with the export countries' requirements for food allergen labelling for ingredients such as celery, mustard or milk. It is the responsibility of the reader to refer to current legislation itself or review the contents of this Guide with the support of a competent adviser, and not to rely on an interpretation or an abridged version of legislative requirements as given in this document.

1.5   Absolute terms, such as 'ensure that', 'avoid', 'prevent', 'absence of' and so on, have been used in various parts of the Guide. To dispense with them would detract from the intentions of the Guide or would necessitate lengthy explanations on each occasion. Accordingly, readers should note that such terms are to be interpreted in a rational and practical way, for example 'ensure that' should be read as meaning 'ensure, so far as is reasonably practicable, that'. Words such as 'should' are used for non-mandatory advice, and the imperative, for example 'must' or 'shall', is reserved for appropriate mandatory requirements.

1.6   Definitions of some of the terms used in this Guide are given in Appendix I. It is appreciated that other definitions may be equally valid or preferred, and the appendix definitions are simply intended to clarify the meanings attributed to a word or phrase when used in the compilation of the Guide.

1.7   The Guide is an advisory document with a list of supporting, supplementary references. The Guide may be particularly useful to students studying food manufacture, to new entrants to management and to general managers in smaller companies who may be responsible for a range of management functions, each of which may be the sole concern of one or more specialist senior managers in a larger company as well as regulatory officers.

1.8   GMP is not a static concept, but an evolutionary, dynamic mechanism by which overall improvements in manufacturing controls can be developed, implemented and maintained.

1.9   The Guide outlines general principles that may already be contained in published guidelines or codes of practice. As appropriate, the Guide will provide references to the original sources that the reader is then advised to consult in full. The Guide will also make reference, where appropriate, to international private standards such as those developed by the Codex Alimentarius Commission.

1.10   The initial adoption of the **EC Official Control of Foodstuffs Directive** and the advent of the **UK Food Safety Act 1990** as well as existing provisions of the **UK Trade Descriptions Act** and the **UK Weights and Measures Act** gave increasing emphasis to the need for a manufacturer to be able to prove that (s)/he did everything necessary to comply with the law. Thus under the **UK Food Safety Act 1990**, and other subsequent legislation, a manufacturer, retailer or importer charged with an offence may enter the legal defence that (s)/he '**took all reasonable precautions and exercised all due diligence to avoid the commission of the offence by the accused or by a person under the control of the accused**'. In this context, it can be considered that 'precautions' are the measures taken and 'diligence' is the activities undertaken to ensure their effective application. The wording puts the onus of proof on the defendant, and both must be proved and the use of the word 'all' implies that 'some' or 'most' will not be enough. What constitutes 'all reasonable precautions and all due diligence' in a particular instance must relate to the nature of the offence and to other related circumstances. Nevertheless in the case of a safety or a 'nature, substance or quality' offence, a manufacturer who can **prove** that (s)/he has diligently installed and appropriately applied all the relevant measures in the Institute of Food Science & Technology (IFST) Guide to Good Manufacturing Practice will stand a very good chance of having a successful defence. It must also be pointed out that a manufacturer who does not employ appropriate technically competent personnel to specify the product formulation, factory processes and procedures to design and control the continuous monitoring of their correct operation and undertake such validation and verification activities cannot be said to have exercised either adequate precautions or adequate diligence and is unlikely to have a successful defence.

1.11   Responsibility for enforcement within the EU varies from country to country. In the UK it is shared between central, devolved and local government bodies. While the making of legislation in the UK is the function of central and devolved government, the enforcement of food law is primarily (but not solely) the responsibility of more than 400 local authorities (LAs) in the UK, and more specifically LA officers. LA officers can be differentiated as being environmental health officers (EHOs) and trading standards officers (TSOs). The Food Standards Agency (FSA) has a statutory requirement, in consort with other government bodies, to protect public health and consumers' interests in relation to food. Since the publication of the last version of the GMP Guide

(Version 6) there has been a policy review with regard to food regulation to move to a more **risk-based approach**. A risk-based approach is well established in UK food regulation, for example the food establishment intervention rating schemes. This trend is also considering the use of public and private regulatory activities to develop a form of co-regulation. This would include utilisation of information from both public enforcement activities (e.g. EHO inspections) and information from private surveillance and verification activities such as third-party audits and product sampling activities.

The roles and responsibilities of all the authorities and organisations in the UK involved in monitoring compliance with, and enforcement of, feed and food law, plant health and feed and food law are set out in the Multi-Annual National Control Plan (MANCP) for the UK. It is a requirement of **Regulation (EC) No. 882/2004** that all EU member states have such a national control plan in place. The MANCP is produced jointly by the FSA and the Department for Food and Rural Affairs (Defra), with contributions from national and devolved agencies. The paragraphs below are a broad overview of the UK legislative framework relating to food manufacture and specific arrangements may be different within a geographical area or industry sector so this should be considered when reading this Guide.

The Framework Agreement on Official Feed and Food Controls by LAs provides the FSA with the processes required to implement its powers under the **Food Standards Act 1999**. This agreement gives structure to the FSA's supervision of LA enforcement work. The Food Law Code of Practice (FLCP) sets out the way LAs should apply food law, and how they should work with food businesses. LAs must follow and implement appropriate provisions of the Code. Practical guidance is also provided as a further help to enforcement officers.

The EHOs and TSOs are authorised by their LAs to enforce food legislation. Once they achieve certain qualifications, detailed under the FLCP, they are authorised to carry out certain tasks and are provided with powers (under the **Food Safety and Hygiene (England) Regulations 2013** and other equivalent UK legislation as amended) to, for example, enter premises, take samples, gather evidence, issue notices and, under certain circumstances, close premises.

Depending on the structure of local government in the area in England and Wales, food visits may be from TSOs to examine labelling, compositional standards and food contaminants, and EHOs to check on food hygiene. However, in Scotland, Northern Ireland and some Welsh and English authorities, EHOs are responsible for all the food legislation, with TSOs responsible for weights and measures checks. Further, in some areas LAs have combined their resources to a single unit which operates over a number of LA areas. It is incumbent on the manufacturer to be aware of the local regulatory framework in the area in which they operate, to have registered their food business and to comply fully with all requirements. Visits to manufacturing sites by LA officers are to ensure compliance with legislation; the frequency of interventions (visits) to a given manufacturing site is determined as previously described by a risk-based approach.

The actual policy and resources allocated to the inspection premises and sampling of product will depend on the individual LA and therefore there are variations in delivery across the country. However, businesses should be able to benefit from a positive relationship with enforcement authorities, receiving detailed written feedback following inspections and receiving results of sampling exercises. Some companies develop a 'Home Authority' or 'Primary Authority' agreement with their LAs. In the UK, the Better Regulation Delivery Office's (BRDO) Primary Authority Scheme gives businesses the right to form a statutory partnership with a single LA that then provides 'robust and reliable advice for other councils to take into account when carrying out inspections or dealing with non-compliance' (see https://www.food.gov.uk/enforcement/enforcework/compliance/primary-auth).

The **European Union (EU) Official Controls Regulation 2017/625** entered into force on 27 April 2017 and replaces Regulation (EC) No. 882/2004 on official controls and other legislation. It becomes applicable over time with the main application date being 14 December 2019.[1] This Regulation addresses official controls and other official activities performed to ensure the application of food and feed law, rules on animal health and welfare, plant health and plant protection products. The Regulation establishes a single legislative framework for the organisation of official controls performed for the verification of compliance with the rules established at EU level or by Member States seeking to apply EU legislation.

1.12 Abbreviations, for example GMP, have been used in the text throughout the Guide, but have been reconfirmed at the start of each chapter in case the chapter is read in isolation and therefore to minimise the number of times that the reader has to refer to the abbreviations list (Appendix II).

[1] http://eur-lex.europa.eu/legal-content/EN/TXT/?uri=CELEX:32017R0625.

2   QUALITY MANAGEMENT SYSTEM

Principle

There should be a comprehensive quality management system (QMS), so designed, documented, implemented, reviewed and continuously improved, and so furnished with personnel, equipment and resources, as to ensure that specifications set to achieve the intended product quality standards are consistently met. The attainment of this quality objective requires the involvement and commitment of all concerned at all stages of manufacture.

Explanatory Note

2.1   A manufacturer has to comply with the legal requirements relevant to the product manufactured, both in the country where manufacture takes place and the countries where the product is destined for export. The term 'manufacturer' is used here not in an abstract sense, but instead referring to the person who has overarching responsibility for the manufacturing operation and for ensuring that good manufacturing practice (GMP) is embedded into the day-to-day operation of the organisation. From a legal perspective, it is this person, for example in United Kingdom (UK) legislation they are termed the 'business operator', that ultimately has responsibility for ensuring all food products are safe and legal. While embracing these legal requirements, she/he or their appointed designate(s) will have also determined the market requirement that she/he aims to meet, and therefore the overall product quality standards the product needs to meet.

These standards can be *intrinsic* (i.e. associated with the innate nature of the food product) or *extrinsic* (i.e. relate to the way the ingredients, or the product as a whole, has been produced and processed often prior to manufacture). Extrinsic standards can include, but are not limited to, farm production standards, worker welfare and other ethical standards, environmental standards and so forth. Thus the product specification that is established as a result embodies both legal requirements (e.g. those of composition, safety, hygiene and labelling) and market requirements (such as product nature, appearance, flavour, texture, presence or absence and quantity of particular nutritional components, nature of pack, pack size, degree of inbuilt convenience, shelf life, presentation, extrinsic standards and price and so forth). While some commercial and marketing considerations affecting the market requirement specification are outside the scope of this Guide, those relating to the principles of design and development of products and processes to comply with that specification are dealt with in Chapter 12. The product and process design, when completed and validated, then becomes a part of the full product specification. Once established it remains permanent until formally changed. All references in this Guide to compliance with product specifications imply compliance with all of

Food & Drink – Good Manufacturing Practice: A Guide to its Responsible Management, Seventh Edition.
The Institute of Food Science & Technology Trust Fund.
© 2018 John Wiley & Sons Ltd. Published 2018 by John Wiley & Sons Ltd.

the foregoing requirements described as being embodied in the term 'specification'.

2.2 In order to achieve the objectives of GMP, it is necessary to have in place:

1. *Quality Assurance:* that is, to design and plan, as relevant, raw material specifications, ingredients formulation, adequate resources such as processing equipment and environment, processing methods and conditions, intermediates specifications, appropriate packaging and labelling specifications, specification for quantity per pack, specifications for management and control procedures, a specified distribution system and cycle, and appropriate storage, handling and preparation instructions, which, taken all together, are capable of resulting in products consistently complying with the product specification and providing confidence that the products will consistently meet the product specification. These will form elements of the quality plan for the product (see 12.9). Furthermore, an effective verification system confirms that the quality plan, and associated quality activities, and the manufacturing operations will consistently deliver products of the specified quality both at a given point in the production process and throughout the product's shelf life.

2. *Effective Manufacturing Operations:* that is, to validate and manage the operational production/distribution practices to ensure that the capability is translated into reality, so that firstly the process itself adheres to specified design parameters and secondly that the resulting products actually do consistently comply with the product specification (see Decision Makers' Summary). This is relevant for quality, legislative and food safety criteria. Within the manufacturing operations there will be steps in the process where it is critical at that point to effectively manage quality. These points are often termed critical quality points (CQPs) and will also be included in the organisation's quality plan(s) with associated criteria, quality limits and monitoring and verification plan. Within the manufacturing operations there will be steps in the process too where it is critical at that point to control food safety. These points are often termed critical control points (CCPs) and will be identified in the organisation's food safety plan with associated preventative (control) measures, critical limits and monitoring and verification plan (see Chapter 3).

3. *Quality Control:* that is, to have in place an effective monitoring system that checks compliance with specified requirements and defines suitable corrective action in the event of 'out-of-control' occurrences.

4. *Food Control:* that is, to have in place an effective management system that ensures food safety, legality, quality and integrity criteria are consistently met (see Decision Makers'

Summary). An effective food control system requires well-qualified and appropriately experienced individuals working in food control management participating in the development and validation of process controls and specifications that address the safety, legality, quality and integrity of food; competent staff and adequate facilities to do all the relevant inspection, sampling and testing of materials, and monitoring of process conditions and relevant aspects of the production environment (including all aspects of hygiene) and management of potential food safety hazards and food integrity threats; verification activities that are developed and implemented by appropriately experienced personnel in order to demonstrate that the food products and the process are under the appropriate level of control; and rapid feedback of information (accompanied where necessary by advice) to manufacturing personnel, thereby enabling prompt adjustment or corrective action to be taken, and enabling processed material to be approved as fit for either further processing or sale, or to be segregated for decision as to appropriate disposition, for example reject, regrade or reprocessing.

Good Manufacturing Practice	2.3	Thus, GMP may be viewed as having two complementary components, namely effective manufacturing operations and food control (see Figure 1).
Food Integrity	2.4	The Elliott Review of 2014 considered the integrity and assurance of food supply networks. The Review was written following the 2013 European horsemeat incident where horsemeat was substituted for beef in a range of products. The report produced as a result of the Review stated that food integrity was not only concerned with the nature, substance and quality and safety of food, but also involved other aspects of food production such as the way food has been 'sourced, procured, and distributed and being honest about those areas to consumers'.[1] The design of food integrity management systems (FIMS) in food manufacturing is in its infancy as this version of the Guide is being written. The writing of a Codex Alimentarius Commission Standard that focuses specifically on food integrity has begun, but its issue will postdate this Guide. Food crime, the design of FIMS, food crime risk assessment, security and countermeasures are addressed in Chapters 5 to 7 of this Guide.
Quality Management Systems	2.5	Many manufacturers will have developed their own QMS, but increasingly are attaining or seeking to attain certification to a private QMS standard. EN ISO 9000:2015 is an international standard concerned with QMS design, and it describes the requirements of a QMS to assure conformance of product and

[1] https://www.gov.uk/government/uploads/system/uploads/attachment_data/file/350726/elliot-review-final-report-july2014.pdf.

production to specified requirements. BS EN ISO 22000:2005 is a complementary standard that addresses food safety management systems (FSMS) and the requirements for any organisation in the food chain. As a result of the increasing globalisation of food production, other private QMS standards have been developed specifically for food manufacturing. These include the British Retail Consortium (BRC) Global Standard for Food Safety. The Global Food Safety Initiative (GFSI) Standard has been developed to benchmark international private standards and both of these documents are referred to within this Guide.

Effective Manufacturing Operations

2.6 An effective manufacturing operation is one where, as appropriate:

(a) the manufacturing process, equipment, activities, precautions and so on are fully specified in advance, and systematically reviewed in light of experience;
(b) the necessary facilities and resources are provided, including:
    (i)   appropriately qualified personnel,
    (ii)   adequate premises and space,
    (iii) suitable equipment and services,
    (iv)   specified materials, including packaging,
    (v)   specified policies and procedures, including cleaning procedures, and
    (vi)   suitable storage and transport;
(c) the relevant written procedures are provided in instructional form and using clear and unambiguous language, and are specifically applicable to the facilities provided;
(d) operators are trained, instructed and motivated to carry out the procedures correctly, and refresher training is undertaken at appropriate intervals;
(e) records are made (whether manually, by recording instruments or both) during all stages of manufacture, which demonstrate that all the processing steps required by the defined procedures were in fact carried out, and that the quantity and quality of product produced were those expected;
(f) records are made and retained in legible and accessible form, which enables the history of the manufacture and distribution of a batch to be traced;
(g) a system is available to withdraw or recall from sale or supply any batch of product, if that should become necessary; and
(h) a review system is in place to consider actual operational performance against proposed performance and drive the implementation of appropriate preventive and corrective action where appropriate.

Quality Control

2.7 Quality control is the function concerned with determining the compliance of finished products with specifications and with the activities ancillary thereto. It includes the undertaking of inspections and tests to determine the degree of compliance with specifications, the examination of process control data and the provision of rapid information and advice leading to corrective action where necessary. It is therefore a 'lag' activity designed to detect product and process failure rather than, in the case of quality assurance activities, to prevent product and process failure in the first place. The term is also used to designate the department responsible for this function within a manufacturing organisation. What is described in this guide in terms of quality control personnel and their activities does not preclude automatic process adjustment by negative feedback from automatic process monitors/recorders, or production operators receiving such information on-screen and then taking appropriate action, provided that they are suitably trained, and that such procedures are written into the quality control system and that any actions undertaken by personnel are recorded.

Food Control   2.8   The Institute of Food Science & Technology (IFST) uses the term 'food control' to describe a comprehensive system that encompasses the QMS, the FIMS and a FSMS based on the principles of hazard analysis critical control point (HACCP). It is vital that the FIMS and the FSMS, and associated prerequisite programmes (PRPs) and countermeasures (see Chapter 7), interlink with aspects of quality assurance and quality control within the QMS that are appropriate to the products and processes involved and the inherent level of food safety and food integrity risk (see Chapters 3 and 6).

In describing the role of the quality manager below, it is recognised that alternative job titles may be used in a particular setting, but it is important for all food manufacturing organisations to distinguish clearly the management roles of quality assurance (failure prevention) and quality control (failure detection), especially where these are in practice managed by the same person. Effective food control requires that, where appropriate:

(a) the quality manager participates, with others as necessary, in the assurance role of development and approval of specifications, liaising with suppliers in agreeing product specifications and service requirements, and the control function of assessing and approving suppliers on the basis of their ability on an ongoing basis to supply reliably in compliance with the specifications;

(b) adequate resources, facilities and staff are available for sampling, inspection, testing and sensory assessment of starting materials (including packaging materials), intermediates and finished products, and for monitoring process and storage conditions and relevant aspects of the production environment (including all aspects of hygiene);

(c) all samples used for inspection and testing are representative of the batch being sampled, and are collected by personnel under the direction of, and examined with methods approved by, the quality manager. The results of such examination(s) need to be formally assessed against the specification by the quality manager or a competent person designated by him/her to undertake the task;

(d) established procedures are developed, validated, implemented and verified that ensure starting materials and all intermediates are approved unconditionally for use, alternatively are rejected or thirdly designated for a treatment that is intended to bring them within specification, in line with any inspection/test results obtained;

(e) there is rapid feedback of information (accompanied, where appropriate, by advice) to manufacturing personnel, enabling prompt adjustment or corrective action to be taken when necessary, and, in the case of raw material, rapid feedback to the purchasing/procurement function;

(f) a positive release procedure exists, where appropriate, whereby batches of finished products are temporarily quarantined until formally released for rectification, or into normal stock, or for distribution. The use, or lack of use, of a positive release procedure should be determined by an appropriate risk assessment process which is formally undertaken, documented and then reviewed at a frequency deemed necessary in light of operational changes or incidents that may occur;

(g) sufficient reference samples of raw materials and packaging, or records of the result of their inspection, where deterioration could occur, should be retained to permit future examination and analysis, if necessary;

(h) sufficient reference samples of finished products are retained for shelf-life tests and to permit future examination and analysis, if necessary;

(i) customer/consumer complaint samples are examined, the causes of defects are investigated where possible, and appropriate measures are advised and implemented for appropriate corrective action to prevent recurrence;

(j) summaries of quality performance data, in an appropriate form, are provided by quality control to operating functions (e.g. general management, production management, purchasing and cost accounting). These summaries may provide input in the determination of food safety and quality objectives for the business whereby data are routinely analysed to determine performance against defined targets and potentially identify areas for improvement;

(k) a direct interest is taken in the activities and quality assurance procedures of the suppliers of raw materials and packaging materials, and close contact is maintained with their quality assurance departments;

(l) ongoing contact is maintained with the relevant enforcement authorities and matters raised by them are investigated and responded to; in the UK the Food Standards Agency (FSA) and the 'Home Authority' will provide useful contacts;

(m) due heed is taken of new developments in food legislation, especially on changes in compositional standards and labelling requirements that may necessitate changes to specifications for raw materials or finished products, and on European Union (EU) and UK Government proposals for future food legislation and in the countries to which the food manufacturer exports; and

(n) the authority and responsibilities of the production management and the quality management functions, respectively, are clearly defined so that there is no misunderstanding (see Chapter 17).

3   HAZARD ANALYSIS CRITICAL CONTROL POINT

Principle

There should be a comprehensive food safety management system (FSMS), so designed, documented, implemented and reviewed, and so furnished with personnel, equipment and resources, as to ensure that critical limits set to achieve the intended food safety standards are not exceeded. The attainment of this food safety objective requires the design, development and implementation of a hazard analysis critical control point (HACCP) system specific to the manufacturing process and the commitment of all staff to its adoption at all stages of manufacture.

Hazard

3.1   A food safety hazard is an agent or material with the potential to cause harm to the consumer. Classic hazard analysis defines three types of food safety hazard: biological (otherwise called microbiological), chemical and physical. This basic classification of food safety hazards needs to be set in the context of emerging hazards and further hazard types being identified in the future which do not fit easily into this classification, which was developed over half a century ago. Food allergens are constituents of a given food, such as inherent proteins, that have the potential to cause an allergic reaction when handled or consumed by an individual who is sensitive to the said agent (see Chapter 8). In some reference and private system standards, allergens are defined as a separate category of food safety hazard whilst in other documents they are included within the category of a biological hazard. *Intrinsic food safety hazards* arise from the product itself, for example fruit stones, fish bones, bone fragments in meat, or as previously described proteins that can cause an allergenic reaction and so forth. *Extrinsic food safety hazards* arise from people, the manufacturing environment, waste and/or other products being manufactured such as glass, metal, wood, ceramic etc.

Hygienic Practice and Prerequisite Programmes

3.2   The 'hygiene package' of five laws adopted by the European Union (EU) in 2004 aimed to merge, harmonise and simplify the complex hygiene requirements that were hitherto contained within 17 EU Directives. The aim was to create a simple, transparent hygiene policy applicable to all food and all food operators together with effective instruments to manage food safety and food safety management throughout the supply chain. The new hygiene law has applied in member states since 1 January 2006.

Good hygienic practice (GHP) is critical to every aspect of good manufacturing practice (GMP), and throughout this Guide it has been treated as a continuous theme and has deliberately not been made the subject of a separate chapter. The Codex Alimentarius Commission (CAC) recommended international code of practice General Principles of Food Hygiene CAC/RCP 1-1969 (2003; Rev 4) lays down the foundation for ensuring GHP,

Food & Drink – Good Manufacturing Practice: A Guide to its Responsible Management, Seventh Edition.
The Institute of Food Science & Technology Trust Fund.
© 2018 John Wiley & Sons Ltd. Published 2018 by John Wiley & Sons Ltd.

and key aspects are addressed in this Guide. The term for prerequisite programmes (PRP) is often used to identify the procedures, policies and protocols that need to be in place within a food organisation before a HACCP plan can be designed and implemented. A number of these requirements are detailed in the previously mentioned CAC/RCP. Examples of PRPs include GHP, GMP, good agricultural practice (GAP), good distribution practice (GDP) etc. These PRPs contain a number of protocols and standards that define best practice for the construction and layout of buildings, premises, workspaces, storage and transport, personal hygiene protocols, premises hygiene and sanitation procedures, maintenance programmes, calibration, training and pest control programmes, procurement procedures and so forth. The ISO/TS 22002-1:2009 *Prerequisite programmes on food safety – Part 1: food manufacturing* specifies requirements for establishing, implementing and maintaining PRP to controlling food safety hazards. The standard was designed to assist organisations seeking to establish, implement and maintain PRP in order to meet the elements of BS EN ISO 22000:2005 *Food safety management systems: Requirements for any organisation in the food chain*. For further details, consult the Campden BRI Guidelines *Food safety plans: principles and basic system requirements* (2016, Guideline 76, ISBN 978090750388). It is essential that the food business operator or their designate is fully conversant with the requirements of PRP establishment, implementation and verification as they underpin the development of food safety management systems (FSMSs) and good integrity management systems (FIMSs). Whilst the senior management team may delegate the day-to-day operations of FSMSs and FIMSs, they ultimately have the responsibility to ensure they are appropriate for the products manufactured and have been suitably, consistently and effectively implemented. The use of HACCP as a risk assessment tool is only the first step to developing an effective FSMS.

HACCP   3.3   With regard to current legislation in the EU, during the design and implementation of manufacturing operations and control procedures, **HACCP principles must be applied as defined in the EU Regulation (EC) No. 852/2004 of the European Parliament and of The Council, in which Regulation 1 requires:**

- **general implementation of procedures based on the HACCP principles, together with the application of good hygiene practice, should reinforce food business operators' responsibility;**
- **guides to good practice are a valuable instrument to aid food business operators at all levels of the food chain with compliance with food hygiene rules and with the application of the [seven] HACCP principles.**

Regulation 2 (a) to (g) defines those [seven] HACCP principles. An EU Regulation has immediate force on the due date in all Member States. Provisions for enforcement and penalties in the

UK are contained in the Food Hygiene (England) Regulations 2005 and similar Regulations for Scotland, Wales and Northern Ireland (as amended).

3.4   It takes more than common sense or business acumen to be able to comply with these legal requirements. In large- and medium-sized food business establishments, it requires suitable numbers of appropriately qualified and experienced personnel. Even in the smallest food business, it is extremely important that the proprietor or some other responsible person has been trained in the principles of food hygiene and food safety, at least to Level 3 standard. There must be senior management commitment to utilising HACCP, which will be implemented through the operation of the FSMS. This means that senior management must commit the resources required to ensure a FSMS is appropriately developed and implemented, and is effective.

Although food safety is the most important factor considered here, the planning and control principles outlined in this chapter are also applicable to preventing or minimising defects during the quality planning process in respect of intrinsic and extrinsic quality attributes too (see Chapter 12).

3.5   The HACCP system and guidelines for its application is published in the *Codex Alimentarius Commission Food Hygiene Basic Texts* (ISBN 9251040214) and identifies seven principles of HACCP:

1. Conduct a hazard analysis. Prepare a list of steps in the process where significant hazards can occur and describe the preventive measures.

2. Identify the critical control points (CCPs) in the process.

3. Establish critical limits for preventive measures associated with each identified CCP.

4. Establish CCP monitoring requirements. Establish procedures for using the results of monitoring to adjust the process and maintain control.

5. Establish corrective actions to be taken when monitoring indicates that there is a deviation from an established critical limit.

6. Establish effective record-keeping procedures that document the HACCP system.

7. Establish procedures for verification that the HACCP system is working correctly.

3.6   In order to undertake HACCP a multi-disciplinary team should be drawn together. The HACCP team needs to contain personnel who have expertise in areas such as production, engineering, quality control, product technology and procurement. The team members need to have relevant practical experience, knowledge of the products and processes within the study and suitable training on how to undertake a HACCP study and the implementation of HACCP principles. At least one member of the team should have formal HACCP training, but all team members need to be trained on how to utilise HACCP principles in assessing how a food product should be manufactured in order to minimise the potential for a food safety incident occurring. The team is also responsible for ongoing review and management of the HACCP system. In the event that external expertise is sourced to assist with either the development or the maintenance of the HACCP system, it is critical that the management team should not delegate responsibility to the external resource. The management of the HACCP system and the development and implementation of the food safety control system remain the responsibility of the manufacturing organisation. The quality of the external expertise should be formally assessed, including the amount of experience in the food industry and the provision of appropriate references from current clients.

3.7 The scope of the HACCP plan(s), that is, the products produced and processes undertaken at the manufacturing site, should be detailed. Relevant information about food products is usually recorded in a product specification. Product specifications should be reviewed to ensure that they contain the relevant information before the start of the HACCP process and if required should be updated. Relevant information includes:

(a) product composition in terms of ingredients, including the origin of ingredients, nature of the item in the case of fruit or vegetables, whether or not the ingredients or the product itself are, or contain allergens;
(b) the physical and chemical attributes of the food, including those that might limit microbial growth, e.g. salt or sugar content, pH or water activity;
(c) packaging type and standards, e.g. gas modified atmosphere, aseptic packaging or vacuum packed;
(d) storage and distribution requirements;
(e) instructions for use;
(f) intended consumer target group, e.g. the general population or a specific group that may be more vulnerable to the food safety hazards being assessed; and
(g) shelf life and nutrition information.

The nature of the treatment and processing of the ingredients and final product undertaken (e.g. cooking or other heat treatment, chlorine washing, blanching, cooling, freezing, metal detection etc.) may also be defined in the product specification, or an alternative document. This information is especially critical where process activities are specifically designed to reduce the likelihood of a food hazard occurring or surviving the processing treatment, for example heat treatment and foreign body detection.

3.8   HACCP is essentially a preventive methodology that needs to be exercised not only within the confines of the in-factory manufacturing process. It should also be applied to the sourcing and intake of the starting materials and packaging materials, and to the post-process packaging, handling and distribution, and indeed, as far as possible, via appropriate storage, preparation and use instructions on the label, as far as the consumer.

3.9   A process flow diagram needs to be developed to identify each step within the manufacturing process. BS EN ISO 22000:2005 describes a flow diagram as a 'schematic and systematic presentation of the sequence of, and interaction of steps' and states that flow diagrams should be prepared for the process(es) and product(s) within the scope of the HACCP or FSMS. Flow diagrams should include, as applicable, the sequence of process steps from intake through each definable stage, to intermediate and finished products, despatch and delivery to the consumer, to stages where reworking, regrading or recycling takes place and where waste is produced, the introduction of packaging and water (whether as an ingredient or a processing aid, e.g. transport), outsourcing of steps or processes, and sequence of events for finished products, semi-finished materials that go into storage awaiting further processing, by-products (e.g. for animal feed) and waste. Recycling or feedback loops should be included on the flow diagram to aid the determination of potential food safety hazards. Process steps should be numbered, again to aid analysis and development of associated food safety control programmes and, where this is applicable, to indicate segregation of areas on the diagram, that is, which process steps occur in low- or high-risk areas (see 3.10 and Chapter 44).

Verifying the flow diagram involves physically walking the flow diagram in the manufacturing premises. The 'walking' of the process flow diagram is important to identify potential hazards that have not been identified in the initial review stages, to determine the degree of implementation of PRPs and preventive (control) measures in practice, to identify areas of potential cross-contamination, to determine holding periods for a product, especially as a result of equipment breakdown and if these could be to the detriment of the product, and to determine whether all process steps (both forward and back in the case of rework loops) have been included in the flow diagram. This verification activity will also aid the determination of realistic food safety hazards. Verification of the site layout plan should also be undertaken at the same time, especially people, product and process flow. Records of flow diagram verification should be maintained and the frequency of verification should also be agreed based on risk assessment, for example those processes that may be seasonal, subject to equipment substitution etc. Reverification activities ensure that any changes to the process flow diagram or the site layout plan have been adequately recorded, and reverification activities should be scheduled at designated intervals.

3.10   As well as the development of a schematic flow diagram that outlines the individual process steps in preparing, storing, manufacturing and despatching the product, further factors should also be considered. A site layout plan for the site as a whole, and the manufacturing areas specifically, should be developed. This plan should cover internal and external areas. It should identify people, product and process flow, especially where there is the potential for delay, rework or recycling. It should also include the availability of and access to utilities such as water, ice and air, especially in the instance where there may be both a potable and non-potable supply of water. This plan should also identify segregation by area, for example allergen control, low-care/high-care areas or low-risk/high-risk areas, temperature-controlled and ambient areas, and areas of the process where the product is fully enclosed or alternatively where the product may be vulnerable to contamination, depending on the products being handled and stored and the nature of the manufacturing process being undertaken. This means that the site layout plan will vary in complexity between, for example, a sandwich manufacturer with multiple zones within their manufacturing process and an ambient long shelf life product manufacturing business where the product arrives on the premises fully enclosed in its final retail packaging.

Personnel facilities such as toilets, changing rooms, smoking areas, rest rooms and location of hand-washing and sanitation points should be included on the site plan (whether internal or external to the main manufacturing building(s). The location of waste systems, drainage systems and cleaning chemical storage should also be identified as well as the flow of waste to external storage. On a complex manufacturing site this may lead to a number of interconnecting site plans being required. The site plan(s) should be used when considering the potential for contamination with extrinsic food safety hazards (see 3.1). Process design should be reviewed to ensure there is the minimum potential for cross-contamination of this nature. This review should also consider external access and the security requirements in terms of product risk, especially where items are stored in external locations.

As well as routes of movement for materials, personnel, reworked material and waste, the site map should also define internal and external access points for personnel, vehicles, visitors, materials and services (see Chapter 7).

3.11   The hazard analysis should consider all realistic potential hazards that could occur at each stage of the manufacturing process and the potential cause. The CAC *Guidelines for the Validation of Food Safety Control Measures* (CAC/GL 69, 2008) state that the control of hazards potentially associated with foods usually involves the application of *control measures* in the food chain, from primary production, through processing, to consumption. The guidelines describe a control measure as any action and activity that can be used to prevent or eliminate a food safety hazard or reduce it to an acceptable level. In this context the terms preventive measure and control measure can be considered to be interchangeable. For a given food safety hazard, there may be one or more control measures that have influence on the likelihood of a hazard occurring and the degree of severity should it occur.

BS EN ISO 22000:2005 states that when selecting control measures, they should be categorised as to whether they are, firstly, elements of the operational PRP(s), that is, procedural or policy based, or, secondly, elements of the HACCP plan, that is, the control measures are product or process based. The standard states that both PRPs and the HACCP plan need to have specific monitoring and verification programmes in place. For each realistic hazard, analysis is required to take account of the severity of the hazard and the likelihood of it occurring and whether elimination or reduction to an acceptable level is critical to ensure food safety. Account should be taken of subsequent stages in the production process and their potential impact on eliminating or reducing the hazard to an acceptable level and hence the impact any deviation is likely to have on the consumer. Realistic food safety hazards are those that are reasonably expected to occur at any step in the process from the raw material, product, people, premises and wider manufacturing facilities.

Existing processes for determining likelihood and severity can be qualitative (Q) based on subjective knowledge of the products and processes, semi-quantitative (SQ), where numbers are assigned to qualitative parameters, or fully quantitative (QRA) as in the use of microbiological risk assessment (MRA) methods.

(a) Traditionally in the Codex Alimentarius guidelines a decision tree tool is used that, through a set sequence of questions, identifies whether a hazard could occur at unacceptable levels or increase to unacceptable levels. However, this approach requires a manufacturing business to be able to quantify what is deemed acceptable. The use of SQ risk assessment matrices is widespread in the food manufacturing industry for determining risk. In SQ risk assessment processes numbers are assigned to qualitative terms for the parameters being used, for example the use of the numbers one, two and three to represent low, medium and high as risk rankings. Commonly in food safety risk assessment the two risk parameters chosen are severity (the degree of harm should a hazard be realized) and likelihood (the probability that the hazard should be realized). Numbers are assigned based on a low, medium or high rating for each and these are multiplied for both criteria to give a risk ranking for that hazard or 'score'. It should be remembered that this is largely a qualitative approach and requires a high level of knowledge and information to be effective. The types of information that could be of value include:

(a) the last risk assessment and rationale for why the hazards were risk ranked as they were by the HACCP team;

(b) historical data and information about known food safety hazards associated with the specific ingredients used or products manufactured;

(c) recognised guidelines and codes of practice related to the products manufactured and processes undertaken;

(d) relevant legislation in the country of manufacture and the countries to which the products could be exported;

(e) market/customer requirements for risk assessment;

(f) influence of target market for the food product(s), e.g. vulnerable groups such as babies and small children, the elderly, the immune-compromised and those who have food allergies; and

(g) the latest scientific literature and research.

When undertaking hazard analysis consideration should be given to realistic biological hazards and the potential for contamination and multiplication and for survival of any heat treatment (e.g. the ability to form spores) and the individual organism's propensity to produce toxins. With chemical and physical hazards again the potential for contamination and the likelihood of being undetected should be considered and where possible eliminated by the controls in the FSMS.

The World Trade Organisation (WTO) Sanitary and Phytosanitary (SPS) Agreement introduced the term 'appropriate level of sanitary or phytosanitary protection' (ALOP), that is, the level of protection deemed appropriate by a country or Member State establishing an SPS measure to protect human, animal or plant life or health within its borders. The 20th edition of the *Procedural Manual of Codex Alimentarius Commission* defined a food safety objective (FSO) as the maximum frequency and/ or concentration of a hazard in a food at the time of consumption that provides or contributes to the ALOP. This definition recognised that the acceptable level of a hazard may vary at different points in the production, supply and consumption of a food product.

Great care must be applied to the decision process of assessing the severity of a hazard and the likelihood of it occurring, particularly in relation to ensuring all of the requisite information and all of the relevant expertise is available (see above).

For food safety hazards where control is deemed to be *critical* at a given point in the supply chain then this process step is termed a CCP for that hazard (HACCP Principle 2). CCPs are those control points in the process where control at that point will eliminate a food safety hazard or control that hazard to an acceptable level. It is important that all those staff working at process steps that are deemed to be CCPs in the manufacturing process are fully aware of this determination and have been sufficiently instructed and trained to be clear on their role in ensuring the food safety hazard in question is adequately controlled and all associated procedures are fully adopted and effectively implemented. Examples of these process points which may be deemed CCPs are sieves, in-line filters, X-ray machines, metal detectors etc.

A decision to incorrectly discard control of a hazard at a specific process step could mean that a substantive CCP is not identified or is now no longer adequately controlled as per HACCP Principle 2. The Codex publication *Principles and Guidelines for the Conduct of Microbiological Risk Assessment CAC/GL-30 (1999)* provides relevant guidance on this.

3.12 Validation is defined in the CAC Guidelines for the Validation of Food Safety Control Measures[1] (CAC/GL 69-2008) as being the obtaining of evidence that a control measure or combination of control measures, if properly implemented, is capable of controlling the hazard to a specified outcome. Validation can also be described as the process of ensuring that the process and procedural controls in place within a manufacturing operation are capable of effectively managing potential food safety hazards should they occur and eliminating them or reducing them to a safe level. Therefore, effective validation of control measures is a critical element of the 'due diligence' defence (see 1.10). Validation is an activity undertaken at pre-FSMS design, implementation and as a post-FSMS implementation activity. Ensuring that the design of the FSMS remains valid over time may require revalidation activities to be undertaken. The process of validation is therefore the assimilation and evaluation of data to demonstrate that the critical limit determined and/or the target level and tolerances defined are appropriate to control the hazard (see HACCP Principle 3). These data can include, but are not limited to, reference to legislation, scientific data, guidelines, codes of practice or technical information, results from validation studies, historical data arising from monitoring and verification activities or data from similar processes, data from mathematical modelling activities and the use of risk assessment models. Risk assessment models, as previously described in 3.11, can be used to determine whether a specific control measure or a combination of control measures, which may be enforced at different stages of manufacture, is capable of consistently controlling a food safety hazard or reducing a food safety hazard to an acceptable level. Revalidation may be required as a result of system or product failure, process or procedural changes, new scientific or regulatory information or evidence of emergent hazards previously unrecognised in the food industry.

The Food Standards Agency (FSA) publication *E. coli O157 – Control of cross-contamination: Guidance for food business operators and enforcement authorities* (2014)[2] stresses the

---

[1] For the *Guidelines for the Validation of Food Safety Control Measures CAC/GL 69-2008*, see http://www.codexalimentarius.net/web/more_info.jsp?id_sta=11022.

[2] For the Food Standards Agency (FSA) publication *E. coli O157 – Control of cross-contamination: Guidance for food business operators and enforcement authorities*, see July 2014 update https://www.food.gov.uk/business-industry/guidancenotes/hygguid/ecoliguide.

importance of not only validating the HACCP plan, but also focusing on validating the control measures in place to ensure, for example, bacterial loading on fresh produce is reduced on receipt, that physical separation of materials is effective and that disinfectants are purchased and used in compliance with validated dilution levels and contact times. The scope of activities required for full validation of a HACCP plan go well beyond the several examples required here. The validation of a HACCP plan at preproduction stages for a new product is essential and routine revalidation activities should be undertaken within the food manufacturing business at a frequency determined by risk assessment.

3.13 As previously described, CCPs should be determined where control is necessary to eliminate or reduce the risk of an unacceptable level of hazard occurring. In determining CCPs, account should be taken of the intended circumstances of use of the product by the customer or consumer. This should include both normal intended use and realistic deviations from this. Intended use could include temperature-controlled storage, cooking or reheating of the food product. Measurable critical limits need to be established at each CCP. These are values that separate acceptability from unacceptability in terms of food safety. Target levels and tolerances may also be set that take into consideration the potential fluctuations within the process and/or provide opportunity to take action before the product is deemed unacceptable (unsafe and therefore rejected).

3.14 At each CCP, a monitoring system must be developed, and appropriate corrective action needs to be determined in the instance that control is lost at a CCP and a target level or critical limit is exceeded (see HACCP Principles 4 and 5). In the instance of a manufacturer using both target levels, designed to give an early warning so the product is brought back under control before it becomes unsafe, and critical limits which if breeched the product is deemed unsafe, two levels of corrective action will be required. In some manufacturers these are described as 'amber' actions and 'red' actions to help to differentiate their severity to staff working at CCPs. Whatever system is adopted by the manufacturer the staff working at the CCP and monitoring activities associated with the CCP must have a clear, unambiguous understanding of the difference between target levels and critical limits.

The activities undertaken in monitoring a CCP should be clearly identified in specific work instructions or similar equivalent document. Personnel working at CCPs should be able to demonstrate their appropriate level of competence. The training undertaken and the formal assessment of competence should be recorded. The corrective actions determined must be capable of bringing both the product and the process back under control, where possible before unsafe food is produced. These actions must ensure that any product or material that may have been produced while the CCP was not in control is suitably identified, controlled and adequately assessed to determine appropriate disposition.

3.15 Records need to be maintained at each CCP to demonstrate that measurements were undertaken on a routine basis to ensure that CCPs are under control (see HACCP Principle 6). In the event of a loss of control at a CCP, the resultant actions taken also need to be recorded. These records form part of the manufacturer's due diligence defence and should demonstrate that only competent personnel have been engaged in CCP operational and monitoring activities. Where records are in electronic form, suitable evidence should be available as to how the checks have been undertaken and how the records have been verified.

The HACCP plan will contain details of the CCP monitoring programme that has been developed. This may include in-line continuous measurement by processing equipment, on-line measurement at designated intervals by staff reading electronic displays and recording information on a separate paper-based form or hand-held device, and off-line measurement at designated intervals where samples of product, packaging and other materials are taken to a workstation or laboratory. It is important to ensure that when undertaking monitoring activities the sample taken is representative of the batch being assessed (see Chapter 35).

Corrective action must be defined in the event of non-compliance being identified at a CCP or when monitoring a PRP. Monitoring of PRP(s) must be formally defined and staff who undertake such monitoring, for example glass and brittle material audits, hygiene audits, glove audits, knife audits, hand-washing standards etc., need to be aware of the compliance standards required and the actions to take in the event of non-compliance. All instances of non-compliance must be recorded along with the associated required corrective action that has been determined, together with evidence that the corrective action has been completed and then verified to ensure it is effective.

3.16 Verification is the activities undertaken in addition to monitoring to determine if the HACCP system is capable of delivering safe food, whether the manufacturing operation is in compliance with the HACCP plan and/or whether the HACCP plan needs modification and review (see HACCP Principle 7). The HACCP plan should be audited and reviewed at least annually to ensure continuing suitability. CCP records should be verified at intervals defined by the manufacturer to ensure that the HACCP system is implemented and effective. Verification should be undertaken by different personnel to those who undertake the monitoring activities prescribed for PRP(s) and/or the HACCP plan. It is important that verification activities should not just address the HACCP plan, but also PRPs and their continued effectiveness. The key tool for verification is the audit (see Chapter 11). The results of internal audits, complaint data, product withdrawal and recall data, and data on service levels as well as internal records of rework or rejection and microbiological and chemical analysis should also be considered in this process. Any trends should be identified, especially where they indicate a loss of control that has not been suitably managed, and a corrective action programme must be implemented. The frequency of verification should be based on risk assessment.

3.17 The FSMS should be reviewed at least annually. In the event of changes to the product (including formulation and recipes), procedures, processes, site plans and people, product and process flow and processing conditions, responsibilities of personnel, supply or composition of raw materials, packaging and ingredients/recipe, consumer use, packaging, storage or distribution activities, emergence of new knowledge, information or data concerning materials, products, processes or processing conditions, emergence of a new food safety hazard, or determination of a change in risk profile, following a food safety incident, withdrawal or recall, or any other factor deemed necessary, a review should be undertaken. In the event of non-conformity, as described in 3.15, a review should also be undertaken. Depending on the characteristics of the product, there is the potential for a new, emergent food safety hazard to occur. In this circumstance, a full review should be undertaken, and this may require a reconsideration of all aspects to ensure that the FSMS is still capable of consistently delivering safe products. For further details, consult Campden BRI *HACCP: A practical guide* (5th edition, 2015, Guideline 42, ISBN 9780907503828) and the Campden BRI HACCP auditing standard (3rd edition, 2015, ISBN 9780907503835).

3.18 In 2016, the European Commission issued a Commission Notice (2016/C 278/01) on the implementation of FSMSs covering prerequisite programs (PRPs) and procedures based on the HACCP principles, including the facilitation/flexibility of the implementation in certain food businesses. The aim of the explanation here is not to give a detailed explanation of its contents, but instead to highlight some of the key themes in terms of how they may affect the implementation of a HACCP system on a manufacturing site. The notice provides clear links between the FSMS, HACCP and PRPs as already outlined in this chapter (see 3.2) and gives further detail on how HACCP can be used to develop a FSMS. More details can be found at http://eur-lex.europa.eu/legal-content/EN/TXT/?uri=uriserv:OJ.C_.2016.278.01.0001.01.ENG.

Hazard Analysis and Operability Study

3.19 HACCP is just one of a number of recognised methods of hazard analysis including failure mode and effects analysis (FMEA) and hazard analysis and operability study (HAZOP). FMEA seeks to identify which failures in an electrical, mechanical or manufacturing process or system can lead to undesirable situations and the means of detection, safeguards that can be implemented and the required actions. HAZOP is a systematic structured approach questioning the sequential stages of a proposed operation/manufacturing process in order to optimise the efficiency and the management of risk. Thus, the application of HAZOP is most effective when used to assess the design of a proposed food-related operation or for existing processes where there are planned modifications to the current design. HAZOP is therefore used to identify both food safety hazards and potential operational issues that could lead to food safety or environmental hazards and also the impact on manufacturing. This approach should result in a system in which as many CCPs as possible have been eliminated, making HACCP during subsequent operations much easier to carry out.

3.20   HAZOP was developed in the 1960s and was a precursor for the development of HACCP as a means of hazard analysis. The HAZOP approach uses guide words and parameters and identifies potential deviations that could lead to problems such as contamination, filter blockage, corners, bends and dead spaces (which might prove difficult to clean effectively), seal or gasket failure, corrosion or stress fractures.

4   FOOD SAFETY CULTURE

Principle

An effective food safety culture, in combination with a comprehensive food safety management system (FSMS) built on hazard analysis critical control point (HACCP) principles, embeds the planned and formal aspects of food safety through the shared values and operational norms of individuals and groups working in the business.

Background

4.1   There is much academic and food industry interest in food safety culture: what it is, how to identify the type of culture that exists within an organisation and how to measure its effectiveness. Following the work of Deming, Juran, Feigenbaum, Ishikawa, Shingo, Crosby and others, quality culture as a topic in itself has been considered within manufacturing businesses for over half a century. This short chapter can only be an introduction to the subject and aims to signpost elements of food safety culture that should be considered within manufacturing operations. Food safety culture as an element of good manufacturing practice (GMP) sits within an overall consideration of organisational culture not only with regard to the scope of the manufacturer's activities, but also the wider supply chain. Again, organisational culture is a topic that has gained much interest in terms of academic, industry and contemporary writing and can only be touched upon in this chapter.

An organisation's formal food safety culture is built upon a set of values and intentions, often formally defined in the mission statement or associated policies. The formal culture is then implemented based on the attitudes and exhibited behaviour of the staff that work for that organisation. These formally espoused values and intentions sit alongside both shared and individual informal values, beliefs and intentions that influence attitude and behaviour and it is the combination of the formal and informal aspects that create the organisation's overall food safety culture. Food safety culture is influenced by, and in turn influences, how the FSMS is adopted and its efficacy.

Food safety culture has to be led from the top, by the person with overall responsibility for the manufacturing business, and be a central factor in all their communications and decision making. The senior management team need to show ownership of food safety and identify the need for accountability at all levels of the organisation and provide the resources and environment to ensure that food safety controls can be adequately and consistently implemented. If staff perceive that it is acceptable within the organisation to consider food safety as being less important within a whole raft of issues that can arise, profitability, allocation of resources, rewarding of individual for actions they have

Food & Drink – Good Manufacturing Practice: A Guide to its Responsible Management, Seventh Edition.
The Institute of Food Science & Technology Trust Fund.
© 2018 John Wiley & Sons Ltd. Published 2018 by John Wiley & Sons Ltd.

taken, operational performance targets, and so forth, then that will potentially lead to a food safety incident. Food safety culture, as with an organisation's quality culture, has been classified in the past as good or bad, positive versus negative and the aim of this chapter is not to state what these terms look like in practice. Instead this chapter has been written to create awareness of food safety culture and its influence on consistently manufacturing safe food for the consumer. The formalised aspects of food safety have been addressed in depth in other chapters of this Guide so it is the informal aspects that are considered here.

Communication and Messaging

4.2 Food safety values are communicated within a manufacturing organisation through information sharing and messages given to, and received from, staff. Some of these communications are visible within the FSMS and others operate at an invisible level in terms of verification activities (see Chapter 11). Visible communications include emails, information sheets, training materials, procedures, work instructions and more formal policies. Training of staff relies both on information and the need to demonstrate relevancy of that information to a given job role or set of responsibilities. Increasing knowledge and greater understanding of risks alone will not necessarily lead to an attitudinal or behavioural response. Individuals may, faced with a series of barriers or constraints, and conflicting messages about business priorities and expectations, simply filter out what they believe is achievable in their work environment or simply comply with the cultural norms of the organisation. In fact, the given formal and informal elements of culture of a manufacturing organisation can either empower individuals or conversely can fail to provide adequate resources to enable adoption of appropriate practices. This can reduce the motivation of staff to improve what they do and minimise food safety risk or just simply encourage staff to comply with prescribed procedures even when they know they could be improved.

Perceived or actual conflict in the prescribed aims and objectives of specific job roles, e.g. between senior management and other levels of management, or between engineering, quality and production staff, means that multiple cultures can exist within a manufacturing organisation and indeed may also flourish, often causing internal conflict and reducing the effectiveness of the FSMS. Griffith's model of food safety culture[1] has three concentric (as in an onion) levels of culture:

- *Level 1 – Food safety climate*: The outermost layer of food business culture that is considered during verification, auditing and inspection activity and is observable. This level of food safety culture can be modified depending on the situation and the internal and external conditions or constraints, e.g. lack of resources, people, or even the presence of the auditor/inspector.

[1]Griffith, C. (2014). *Developing and Maintaining a Positive Food Safety Culture*. Highfield Publications.

- *Level 2 – Underpinning culture*: The middle layer includes the organisation's espoused values (often unspoken) and guides the employees' behaviour and attitudes to authority and legislation. Depending on the depth of verification activity, this level of culture can be examined and measured. (The depth of audits is considered more fully elsewhere in the Guide, see 11.8.)
- *Level 3 – Core culture*: The innermost layer that contains all the beliefs and assumptions by staff as individuals or groups about what the organisation stands for. This level includes core values that are invisible and often assumed. Depending on the depth and scope of any verification activity this level may remain hidden (see 11.8).

Each manufacturing business will be different in terms of the exact combination of these levels of culture and thus will have a unique food safety culture that influences the degree of effectiveness of the FSMS. The type and quality of the verification activity will determine the level of understanding of the overarching food safety culture in the manufacturing organisation and the types of subcultures that may exist and the impact of this dynamic on the effectiveness of the FSMS.

4.3 The link between the effectiveness of the FSMS and the associated food safety culture of the organisation is clear. The British Retail Consortium (BRC) Standard for Food Safety (Issue 7) states that it is through the use of unannounced audits that there can be greater confidence in the food safety culture of the manufacturing organisation. In partnership with Campden BRI and Taylor Shannon International (TSI) the BRC have developed a food safety culture module adopted alongside the annual BRC Global Standard for Food Safety verification audit. This is just one of a growing number of food safety culture assessment tools.

Food Safety Culture Assessment Tools

4.4 A number of tools have been developed to assess food safety culture, including food safety culture questionnaires, behavioural observation and the development of tools that underpin system review and the development of performance indicators. Assessment tools such as the BRC Food Safety Culture module are based on two questionnaires, one completed by staff and the other by a third-party auditor. As a result the site will receive a score based on the overarching food safety culture and then supplementary categories: people, process, purpose and proactivity. A manufacturing organisation is graded based on the score achieved as a result of the assessment. An action plan can then be implemented to address weaknesses and areas where corrective action is required.

5 FOOD CRIME AND FOOD INTEGRITY MANAGEMENT SYSTEMS

Principle

Crime can arise either internally within the manufacturing business or externally from those organisations with which it has a commercial interface. In order for the food business operator to understand the degree of food crime risk associated with the products they produce and the ingredients and materials they source, they need to be aware of the potential for criminal activity both within and outside of their organisation. Food business operations need to also consider the wider social and environmental factors that can lead to illicit activity in the food supply chain. Food crime extends beyond the legislation associated with food safety and food labelling to encompass all activities associated with food manufacture and supply.

Introduction

5.1   The United Kingdom (UK) Elliott Review of 2014, written following the 2013 European horsemeat incident, considered the integrity and assurance of food supply networks (see 2.4), stated that food integrity was not only concerned with the nature, substance and quality and safety of food, but also involved other aspects of food production such as the way food has been 'sourced, procured, and distributed and being honest about those areas to consumers'.[1] The design of food integrity management systems (FIMSs) in food manufacturing is in its infancy as this version of the Guide is being written. The writing of a Codex Alimentarius Commission Standard that focuses specifically on food integrity has begun, but its issue will postdate this Guide.

The concept of food crime is not new. The National Food Crime Unit (NFCU) formed as a result of the recommendations of the Elliott Review uses the following definitions:

Food fraud: A dishonest act or omission relating to the production or supply of food that is intended for personal gain or to cause loss to another party; and

Food crime: Dishonesty relating to the production or supply of food that is either complex or likely to be seriously detrimental to consumers, businesses or the overall public interest.[2]

The NFCU state that food fraud becomes food crime when the scale and potential impact of the criminal activity is considered to be serious in terms of geographic impact (cross-regional, national or international reach), there is significant risk to public

---

[1]https://www.gov.uk/government/uploads/system/uploads/attachment_data/file/350726/elliot-review-final-report-july2014.pdf.
[2]https://www.food.gov.uk/enforcement/the-national-food-crime-unit/what-is-food-crime-and-food-fraud.

Food & Drink – Good Manufacturing Practice: A Guide to its Responsible Management, Seventh Edition.
The Institute of Food Science & Technology Trust Fund.
© 2018 John Wiley & Sons Ltd. Published 2018 by John Wiley & Sons Ltd.

safety or there is a substantial financial loss to consumers or businesses. For more information on the NFCU see the Food Standards Agency (FSA) website: https://www.food.gov.uk/enforcement/the-national-food-crime-unit.

5.2 The term 'food fraud' has been developed (by John Spink and others) to encompass a range of criminal activity, including:

- *misbranding or misrepresentation* on labelling, packaging and websites and misleading indications, e.g. by using words and pictures that do not reflect the actual production methods employed to produce the food, incorrect use of the words such as 'traditional' and 'natural', the intentional sale of eggs from caged reared birds substituted as free range and/or the use of oversized packaging that misrepresents the size of the food portion inside;
- *counterfeiting*, where the product and packaging is fully replicated to look like the original product and sold on as such;
- *simulation*, where illegitimate product is designed to look like, but not be an exact copy of, the legitimate product, which could be as a result of poor replication of labelling on an illegitimate product;
- *overrun* where it is illegitimately used; overrun is a product made in excess of production agreements, which may be a concern for a manufacturer when outsourcing production to another organisation, or where overrun is bought and the indication of minimum durability is altered without sanction by the original manufacturer;
- *diverting*, where product is sold through non-prescribed distribution channels. An example of this would be where following overproduction of food products (overrun) these additional products are packed in their original retailer packaging then sold off within that packaging to an alternative customer, e.g. for sale in a street market or car boot sale when the product itself was destined for other markets;
- *tampering* with malicious intent for personal gain. The potential for this type of food crime has been understood and procedures put in place to mitigate for it for several decades, e.g. the use of tamper-evident packaging or tamper-evident seals being placed on transport containers and vehicles;
- *substituting* one ingredient for another less expensive alternative. This has been a frequent occurrence in history, especially if such activity can be undertaken whilst unnoticed by purchasers and ultimately the consumer. Examples include substituting sawdust in bread in the 1860s, putting antifreeze in wine in the 1980s, or in the recent decade melamine substitution in milk and gluten, white fish substitution for cod or haddock and horsemeat for beef. Intentional substitution of a food product can cause harm, e.g. the undisclosed substitution of peanut powder in almond powder, a risk for those with peanut allergy. However, substitution can be a lucrative activity as it may well go unnoticed if the agent is not harmful to the

consumer. This is especially so where there is a lack of awareness in the food industry that the substituting agent in question might be used and thus no product testing is undertaken to determine its potential inclusion. Where the substituting material is neither harmful nor readily tested for, this can be seen by the unscrupulous as a ripe opportunity for financial gain. Substitution is a global issue and with the materials and ingredients they use no manufacturer is without risk in this regard. There is potential for inadvertent substitution as a result of human error or the implementation of lax manufacturing procedures such as the failure to sign off recipe sheets when items are included, the failure to undertake visual checks of ingredients before inclusion, poor labelling of visually similar ingredients in store, weak control of decanted ingredients and a failure to adequately label/identify the material when it is in alternative packaging, and the failure to undertake stock control activities in ingredient stores to determine if the incorrect ingredients have been used before products are released to distribution and so forth. Thus substitution may or may not be intentional, but it is illegal to sell a product that does not comply with the associated labelling information.

- *adulterating* by intentionally adding extraneous, improper or inferior ingredients to a food product. The term 'economically motivated adulteration' (EMA) is used to describe the process of adulteration with the specific intent of reducing the cost of production or misrepresenting a food to increase profits. The scope for intentional adulteration is vast within the food sector and each manufacturing organisation needs to undertake a focused risk assessment to determine both the ingredients that they purchase and the products that they manufacture that are at risk of adulteration. This risk assessment should also be undertaken for all outsourced activities (see Chapter 33). An appropriate source for up-to-date information on food adulteration in the UK is the Food Authenticity Network (see http://www.foodauthenticity. uk). The European Union (EU) food and feed alert portal (RASFF) is a useful source for emerging and re-emerging instances of food adulteration identified in European member states (see https://ec.europa.eu/food/safety/rasff/portal_en) and the United States (US) Food Adulteration Incidents Registry (FAIR) compiles details of both EMA and food defense incidents (see https://foodprotection.umn.edu/fair); and
- simple *theft* from food premises or the wider supply chain.[3]

5.3   The foods most vulnerable[4] to food crime include those that are high-value ingredients, such as spices, olive oil, fish, honey, wine, coffee etc. where substituting by an alternative ingredient can provide significant financial gain. Foods associated with an

---

[3]Spink, J. and Moyer, D.C. (2013). Understanding and combating food fraud. *Food Technology*, 67(1): 30–35.
[4]WHO (World Health Organization) (2008). *Terrorist threats to food. Guidance for establishing and strengthening prevention and response systems.* Food Safety Issues Series. WHO, Geneva.

ideological, ethnic or religious grouping are also vulnerable, especially if that food can be seen as a representation of that group itself, e.g. kosher or halal products. Intentional ideological food crime of this nature is often termed a food defence issue (see 5.4). Foods are also vulnerable where the supply chain crosses a number of borders or where activities by food criminals could go undetected. Examples of this include foods disseminated or distributed widely though multiple businesses with complex interfaces, foods produced, manufactured and stored in readily accessible or poorly supervised areas or where staff working in those areas have little awareness of the potential for food crime (see Chapter 7), and foods susceptible to tampering or interference where this can go undetected by the inspection, testing, monitoring and verification activities routinely undertaken in the food supply chain.

Food Defence	5.4	Food defence is a term used to describe the activities required to protect a food product or indeed a food supply chain from intentional or deliberate acts of contamination and includes ideological or malicious threat. The Global Food Safety Initiative (2013) describes food defence as 'the process to ensure the security of food and drink and their supply chains from all forms of intentional malicious attack including ideologically motivated attack leading to contamination or supply failure'.[5] Thus food defence strategies can be developed at the national, regional, supply chain and organisational level. The US Food and Drug Administration (FDA) has differentiated between national risk assessment models and supply chain or organisational food defense models[6] (see Chapter 6).
Cybersecurity	5.5	*Cybersecurity* in its simplest sense is the measures taken to protect a computer system or individual appliance against an intentional malicious target attack and/or unauthorised access and unintentional or accidental access. These measures include developing policies and procedures, undertaking relevant risk assessment approaches, undertaking training and awareness sessions commensurate with an individual staff member's responsibilities and developing soft or hard controls such as specific software, firewalls, technologies etc. that can protect the organisation's cyber environment and their electronic assets. This protection includes not only the prevention of damage, but also the ability to restore computers and electronic communication systems. ISO/IEC 27032 (2012) *Information technology – Security techniques – Guidelines for cybersecurity* defines cybersecurity risks as including social engineering attacks, hacking, the proliferation of malicious software ('malware'), spyware and other potentially unwanted software. The technical guidance suggests that

---

[5]GFSI (2013). *The Global Food Safety Initiative GFSI Guidance Document*. Version 6.3, October.
[6] *Vulnerability assessment software*. Available from http://www.fda.gov/Food/FoodDefense/ToolsEducational Materials/ucm295900.htm#whatmean.

measures should be adopted to address these risks, including controls for 'preparing for attacks by, for example, malware, individual miscreants, or criminal organisations on the Internet; detecting and monitoring attacks; and responding to attacks'. Electronic assets include formulae, recipes and specifications for products, electronic controls for key equipment that support the maintenance of food safety systems, and other essential data.

Threat Analysis   5.6   PAS 96:2017[7] differentiates between accidental contamination of food, which is addressed by hazard analysis critical control point (HACCP) plans (see Chapter 3), and threat assessment critical control point (TACCP), which considers deliberate or intentional acts that affect the integrity of the food and the food supply chain (see Chapter 6). Whilst a hazard is considered to be any agent with the potential to cause loss or harm which arises from a naturally occurring or accidental event or results from incompetence or ignorance of the people involved, a 'threat' is defined by PAS 96:2017 as 'something that can cause loss or harm which arises from the ill-intent of people'.

PAS 96:2017 identifies the steps that need to be taken to develop food protection and defense procedures and an associated emergency response and business continuity management system (BCMS) to ensure that the business can effectively manage and minimise the disruption caused by a potential incident (see Chapter 27 for more details on BCMS). BS 11200:2014 *Crisis management: Guidance and good practice* gives guidance to help plan, establish, operate, maintain, monitor, verify and review an organisation's crisis management plan and its wider capability (see Chapter 27). Business continuity plans are an important element of the manufacturer's formal management systems. They could be situated within the quality management system (QMS), food safety management system (FSMS) or food integrity management system (FIMS) or an integrated overall management system that covers all areas. In the event of suspected, or actual, illegal activity associated with a food item these plans will ensure the continuity of supply to customers should an incident occur that disrupts 'normal' manufacturing activity. The type of potential incident that could occur on the manufacturing site needs to be determined so that a threat analysis assessment can be undertaken using assessment tools such as TACCP similarly to how HACCP might be used as a management tool to undertake food safety hazard analysis.

Types of Food Criminals   5.7   In order to develop a robust, reactive FIMS that is appropriate to the business and the food products it supplies, an understanding of the types of food criminal that might be at work and their motivations is crucial. Food criminals[8] have been divided into

[7] https://www.food.gov.uk/sites/default/files/pas962017.pdf.
[8] Spink, J. and Moyer, D.C. (2013). Understanding and combating food fraud. *Food Technology*, 67(1): 30–35.

several categories here using the work of Manning et al. (2016).[9] *Recreational criminals* undertake crime for its entertainment value. This type of food crime can be mitigated against by implementing appropriate security measures so that the crime becomes harder to perpetrate and then is of higher risk to the criminal in that manufacturing setting compared to other illicit activities they could undertake. *Occasional criminals* commit crimes infrequently or are opportunistic. This category of criminal can be stopped by implementing simple security procedures at the manufacturing site, e.g. addressing any security doors left open or unknown personnel being on site without being challenged etc., and developing an associated training plan for staff (see Chapter 7).

Occupational criminals commit crime at their place of employment. They may be acting alone, collaborating with others in the organisation or working as part of a wider supply chain network. There are two types of occupational criminal. The first is the type of perpetrator who infiltrates an otherwise legitimate business with the intention that they can then undertake their illicit activities for personal gain. Some individuals may engage in 'wilful blindness' about the activities occurring in their own business or other businesses that they engage with, often for their own financial gain. This type of food crime is a major risk to a legitimate manufacturing operation, especially if the individual or their collaborators are in positions within the organisation where their activities can be covered up and can go unnoticed. The second type of occupational criminal is the one that perceives a business culture that accepts illegal activity as part of everyday operations in order to meet the demands and pressures of the wider supply chain. The business culture that is promoted and championed by all levels of management within the manufacturing organisation must be seen at all times to completely condone the taking of shortcuts or trade-offs that allow non-compliant or non-standard product to be despatched, and/or managers deliberately ignoring activities that may be taking place perpetrated by staff in the name of benefiting the organisation, but that in truth are ethically or morally inappropriate or even illegal.

Professional criminals are individuals whose total lifestyle is financed by criminal and illegal activity. They can interact with a manufacturing business with the sole objective of personal financial gain. The legal business may, or may not, be aware of the professional criminals' illegal activities. With professional criminals, illegal trading in food and food ingredients may only be an element of their overall business activity.

Ideological criminals are those individuals or collaborators who commit criminal acts in order to make an ideological or political

---

[9]Manning, L., Smith, R. and Soon, J.M. (2016) Developing an organizational typology of criminals in the meat supply chain. *Food Policy*, 59: 44–54.

statement, to cause economic harm to a specific business, region or country with which they have a personal grievance or seek to create fear or panic in a target population. These individuals may be focused on sabotage, contamination with toxic materials, extortion and blackmail, bioterrorism or other forms of malicious activity.

PAS 96:2017 *Guide to protecting and defending food and drink from deliberate attack* has its own typology of food criminals and describes them as the extortionist, the extremist, the opportunist, the hacktivist and other cyber criminals (see 5.5), the irrational individual, the disgruntled individual, and the professional criminal. Thus in order to determine a particular threat it is important to consider the motivations of the perpetrator.

Food Integrity Management Systems

5.8 The category of food fraud or food crime and the type of food criminal involved will influence the good manufacturing practice (GMP) controls that need to be adopted to mitigate against the potential for their occurrence. Food crime mitigation through designing, implementing and verifying a FIMS may be considered by many as a new and emerging element of GMP. Whilst historically food crime mitigation has not been explicitly identified as an area of food control and GMP, there has been an implicit requirement to ensure that food is safe, legal and of the required quality. However, this chapter and Chapters 6 and 7, which are new to this edition of the Guide, have been written with specific emphasis on the practical measures that food business operators can undertake to minimise their risk of being victims of food crime. As already outlined in this chapter, the requirement to undertake risk assessment activities in order to inform an effective FSMS is equally the case with the development of a FIMS. As a result, food crime risk assessment models have been developed and continue to be redeveloped so that business operators and retailers can use them to address the potential for food crime and how it can be mitigated (see Chapter 6).

The food industry has developed and adopted many operational prerequisite programmes (PRP), i.e. procedures or protocols to minimise food safety risk and potential harm to the consumer (see 3.2). These PRPs, such as good hygienic practice (GHP) and GMP, together with food inspection techniques will also have a role within a FIMS. Indeed the way to manage food safety, legality and quality issues is often through an integrated approach where the FSMS, FIMS and QMS are formalised together with additional management systems to address personnel health and safety and environmental management in one documented system for the manufacturing organisation. This integrated approach ultimately leads to a streamlined documented system.

5.9 Product identification and traceability (see Chapter 14) and provenance and authenticity (see Chapter 15) can be difficult to maintain in complex global chains, especially where they include national or regional borders over which regulatory requirements and inspection regimes may vary. This makes such chains open to fraud. There is a greater food integrity risk associated with food chains with minimal supplier–customer relationships that are often lacking in transparency, with minimal traceability controls in place and where little or no assurance activity is practiced. Food manufacturers must as part of their food integrity risk assessment activities identify the individual food products and materials that they procure, supply and/or produce which fall into the category of foods vulnerable to food crime (see Chapter 6). Some organisations have developed a risk assessment approach to procurement in terms of assessing vulnerability in order to map their supply chain vulnerability and the degree of risk (see the Elliott Review in Appendix J[10] for examples).

5.10   In order to develop an effective FIMS it is critical that the manufacturer understands and can communicate the organisation's needs and requirements for establishing a food integrity policy and food integrity objectives that guide the overall FIMS. It is important for the manufacturer to assess the threats and associated risks appropriately that relate to food integrity management, to develop, implement, monitor, review and verify food integrity processes, controls and measures designed to treat and mitigate food integrity risk, and to implement a programme to drive continual improvement in terms of both known and emerging threats.

5.11   A FIMS, in line with FSMS and QMS, needs to include the following elements: a food integrity policy or an integrated organisational policy that addresses food integrity, defined responsibilities for the implementation and verification of the processes designed to treat and mitigate food integrity risk, a clear operating plan that addresses food integrity, a mechanism to ensure awareness of food integrity issues and a training programme that is designed to ensure that staff are competent and have the knowledge and skills required to implement their responsibilities consistently and effectively. There needs to be a formal management review and performance assessment protocol for the FIMS, which is adopted in line with requirements and is appropriate for the threats that are associated with the products and the manufacturing site. Adequate resources must be in place to ensure that the prescribed requirements of the FIMS can be effectively implemented. There must be clear, unambiguous documentation to support the activities required to implement the FIMS and demonstrate its efficacy.

5.12   The management team must develop a formal policy to ensure that the need for food integrity management is adequately

[10]https://www.gov.uk/government/uploads/system/uploads/attachment_data/file/350726/elliot-review-final-report-july2014.pdf.

addressed and suitable controls are developed and referred to within the FIMS. This policy should define the organisation's intentions and commitment to supplying legal products and should be signed by the senior management individual with overall responsibility for the safety, legality and quality of the products manufactured. The organisation must demonstrate that it has communicated this policy to all personnel and that they understand its importance. Ongoing refresher and update training must be undertaken with staff, especially when the organisation is aware of supply chain information (intelligence) that may suggest the organisation or the products that they produce could be a potential target for food crime (see Chapter 17). Consideration should be given to cultural differences, and reading or language difficulties of staff when these objectives are communicated.

5.13   The organisation's strategic policy on food integrity must interface with specific customer (including retailer) code of practice or supplier requirements with regard to food legality, procedures for whistleblowing or other relevant security procedures. For more details on security and countermeasures see Chapter 7.

5.14   The ability of the organisation to effectively manage food integrity should provide an input into the management review and internal audit and verification process (see Chapter 11). The management review and associated verification activities undertaken should determine the degree of management control within the FIMS and its effectiveness, and verify that the organisational objectives associated with food integrity are suitable, realistic and consistently being met. It is also an opportunity to identify any areas for improvement. An internal audit programme should include food integrity audits and associated verification activities. These need not be stand-alone audits. Instead, the scope of the existing audit programme should be extended to include specific types of food integrity audits in the context of food safety, legality and quality.

6   FOOD CRIME RISK ASSESSMENT

Principle

Risk assessment and, more specifically, food safety risk assessment approaches such as hazard analysis critical control point (HACCP) have been used in food manufacturing for over 50 years. Food crime risk assessment (FCRA), however, is a much newer phenomenon and many food manufacturers are developing their own risk assessment approaches and associated food integrity management systems (FIMSs). The mechanisms for FCRA are outlined in this chapter along with how the use of risk management tools to determine organisational vulnerability can then inform appropriate countermeasures and security controls to protect food integrity.

Introduction

6.1   BS EN ISO 31000:2009 *Risk management – Principles and guidelines* is an international standard for risk management processes and the contents are of value for manufacturers in terms of all aspects of risk assessment. The standard BS EN ISO 31000:2009 provides within the standard principles a framework and a process for managing risk. Risk assessment is influenced by the quality of knowledge and information available on which to base any given risk assessment and also the degree of certainty (or uncertainty) on whether any given event, consequence or likelihood can be adequately determined. BS EN ISO 31000:2009 provides a hierarchy of how risk should be dealt with:

1. avoiding the risk by deciding not to start or continue with the activity that gives rise to the risk;

2. accepting or increasing the risk in order to pursue an opportunity;

3. removing the risk source;

4. changing the likelihood;

5. changing the consequences;

6. sharing the risk with another party or parties (including contracts and risk financing); and

7. retaining the risk by informed decision.

An understanding of the types of food criminal and their motivations is crucial when undertaking a formal FCRA and then when developing a FIMS. The PAS 96:2017 *Guide to protecting and defending food and drink from deliberate attack* has its own typology of food criminals and other typologies have been developed too (see 5.7).

Food & Drink – Good Manufacturing Practice: A Guide to its Responsible Management, Seventh Edition.
The Institute of Food Science & Technology Trust Fund.
© 2018 John Wiley & Sons Ltd. Published 2018 by John Wiley & Sons Ltd.

6.2 Food manufacturers, as part of their FCRA process, must firstly identify the individual food products that they procure, supply and/or produce which fall into the category of foods that are vulnerable to food crime (see 5.3). *Vulnerabilities* are the weak points or gaps in the formal management systems, or on the manufacturing site itself, that can be identified by perpetrators where their intentional action to mislead, misinform and/or undertake illegal activity can lead to a realisable threat (see 5.6). These vulnerabilities or weak areas within manufacturing operations or in the wider supply chain are often termed *hotspots*. Vulnerability risk assessment is the process that determines the weaknesses or exposed elements of a manufacturing organisation or supply chain where illicit activity can take place in order to determine how susceptible the system is to that potential threat and to mitigate such weaknesses where they are identified so appropriate action can be taken to reduce risk. Vulnerability will be reduced if there is a high level of deterrence and this will depend on the security measures and types of countermeasures that are in place (see Chapter 7). Vulnerability risk assessment tools can be applied at national, supply chain and individual business levels and these are now explored in this chapter.

Threat Assessment 6.3 PAS 96: 2017[1] differentiates between accidental contamination *Critical Control Point* of food, which is addressed by HACCP plans (see Chapter 3), and intentional contamination. A *threat* is defined by PAS 96:2017 *Guide to protecting and defending food and drink from deliberate attack* as 'something that can cause loss or harm, which arises from the ill-intent of people'. PAS 96:2017 identifies generic threats to food and drink including economically motivated adulteration, malicious contamination with toxic materials causing ill-health and even death, sabotage of the supply chain, leading to food shortage, and misuse of food and drink materials for terrorist or criminal purposes. Threat analysis critical control point (TACCP), as outlined in PAS 96:2017, is a risk management methodology that aligns with HACCP but focuses on deliberate or intentional acts that affect the integrity of food products themselves and the wider food supply chain. PAS 96:2017 describes TACCP as a 'systematic management of risk through the evaluation of threats, identification of vulnerabilities, and implementation of controls to materials and products, purchasing, processes, premises, distribution networks and business systems by a knowledgeable and trusted team with the authority to implement changes to procedures'. TACCP aims to:

- reduce the likelihood (chance) of a deliberate attack;
- reduce the consequences (impact) of an attack;
- protect organisational reputation;
- reassure customers, press and the public that proportionate steps are in place to protect food;

---

[1]https://www.food.gov.uk/sites/default/files/pas962017.pdf.

- satisfy international expectations and support the work of trading partners; and
- demonstrate that reasonable precautions are taken and due diligence is exercised in protecting food.

For the manufacturer and in comparison with HACCP, PAS 96:2017 states that TACCP should be a team approach that follows 15 specific steps:

1. Assess new information – the team should evaluate all emerging information (intelligence).

2. Identify and assess threats to the organisation – identify individuals and/or groups which may be a threat to the organisation and assess their motivation, capability and determination.

3. Identify and assess threats to the operation – identify individuals and/or groups which may be a threat to the specific operation (e.g. premises, factory, site).

4. Select product – that is representative of a given manufacturing process.

5. Identify and assess threats to the product – identify individuals and/or groups that may want to target the specific product.

6. Devise a flow chart of the product supply chain – consider all aspects, especially those that may be less transparent.

7. Identify key staff and vulnerable points – for each process step in the flow chart identify the vulnerable points where an attacker might target and the people who would have access to the product at that point.

8. Consider the impact of threats identified that are appropriate to the product and assess the impact the process may have in mitigating the threat.

9. Identify which supply points are most critical – either where the threat might have the most effect and/or where the threat might be best detected.

10. Determine if control procedures will detect the threat – assess the likelihood of routine control procedures detecting such a threat.

11. Prioritise based on likelihood versus impact – using a scoring system score the likelihood of the threat and the impact that the threat might have for all products. This will assist in prioritising the threats. The suggested TACCP scoring system is semi-quantitative, i.e. associates a score (number) to a qualitative statement both for impact and likelihood. A risk matrix approach can then be used to highlight overall summative risk. There is an assumption in this approach that both the characteristics of impact (minor, some, significant, major and catastrophic) and likelihood of threat (unlikely to happen, may happen, some chance, high chance, very high

chance) can be determined, i.e. are known and that there is consistency between individuals undertaking risk assessment in how they derive the qualitative term and the associated score. Impact and likelihood can be reduced by the adoption of appropriate countermeasures (see Chapter 7).

12. Identify who could carry the threat out in practice – if the priority is high consider the individuals who have unsupervised access to the product and process and their personal integrity.

13. Decide and implement necessary controls – identify, record confidentially, agree and implement proportionate preventative action (critical controls). Confidentiality is crucial so that weak spots are not readily identified to potential perpetrators (attackers).

14. Review and revise the TACCP plan.

15. Monitor by horizon scanning and identification of emerging risks (return to step 1) – develop an early warning system that can highlight new threats or a change to existing threats.

The term *horizon scanning* is increasingly being used to consider the supply chain and a given food operation's food crime vulnerability and the *countermeasures* that need to be introduced to ensure effective control. Horizon scanning is systematic in nature and considers the existing information, evidence or intelligence available about products, processes and the wider supply chain as well as socio-economic factors that could influence future trends and scenarios with regard to food crime. Effective horizon scanning considers evidence from this range of sources in order to effectively map potential threats and vulnerabilities and then identify the potential for their occurrence and the means for their control. Horizon scanning is not an on–off exercise. Horizon scanning needs to be a dynamic management approach that can react to changes in the supply chain, with products and ingredients, and within the manufacturing operation itself. It is important to have formalised mechanisms for updating horizon scanning assessments if the evidence base changes in the future and as a result the ranking for a given threat and associated risk changes.

Vulnerability Risk Assessments   6.4   There are a number of methods that have been developed to determine vulnerability as part of a horizon scanning process. These include emerging risk determination, global chain analysis and incident classification matrices.

Emerging risk determination is a systematic approach that uses a protocol, intelligence strategy and expert knowledge to detect emerging risk utilising a set of questions namely:[2,3]

[2] https://www.food.gov.uk/sites/default/files/multimedia/pdfs/board/fsa110109.pdf.
[3] https://www.food.gov.uk/sites/default/files/multimedia/pdfs/terrydonohoepres.pdf.

What is typical? Identify and characterise the current status within the food chain to form a baseline.

What is exceptional? Use available intelligence to monitor movements against the baseline and identify variance.

How do we prevent recurrence? Determine root causes of reoccurring issues so corrective action is targeted and effective.

What don't we know? Analyse global food chains and identify vulnerabilities that could lead to future threats.

Global chain analysis (GCA) determines vulnerability and the risk of illicit behaviour at the stages in the food supply chain where food crime could occur.

Incident classification matrices take a risk assessment approach to determine whether a given food crime incident is classified as low, medium or high risk.[4] The method uses a scoring system based on the criteria of severity (health effects, consumers affected, risk assessment, perceived risk by consumers and the media) and complexity (number of reports, number of products, number of agencies involved, traceability).

These three models operate at the supply chain level as opposed to the organisational level, but a number of risk assessment approaches use a matrix approach at the level of the manufacturing unit.

6.5   The United States (US) Food and Drug Administration (FDA) distinguished between national risk assessment models and supply chain or organisational food defence models.[5] The model that was developed considers vulnerability specifically with regard to intentional ideological contamination. In the US, the CARVER + Shock method was adopted, where the acronym CARVER stands for:

Criticality: a measure of the public health and economic impacts of an attack as a result of the batch size or network of distribution;

Accessibility: the ability to gain physical access and egress where this can change over time and also as a result of the use of countermeasures;

Recuperability: the ability of a food system to recover from an attack;

[4] http://www.food.gov.uk/multimedia/pdfs/incident-response-protocol.pdf.
[5] Vulnerability assessment software. Available from http://www.fda.gov/Food/ FoodDefense/ToolsEducational Materials/ucm295900.htm#whatmean.

Vulnerability: the ease of accomplishing the attack, which can change over time and as a result of the use of countermeasures;

Effect: the amount of direct loss from an attack as measured by loss in production;

Recognisability: the ease of identifying the target, with

Shock, a combined measure of the health, psychological and collateral national economic impacts of a successful attack on the target system, being the final element.[6]

Risk is determined via a weighted score-based approach. The FDA and the US Department of Agriculture (USDA) adapted the CARVER + Shock approach to develop facility or process level food defence vulnerability assessment using a vulnerability assessment software (VAS) tool to develop a focused food defence plan for the given operation. In so doing, the VAS tool combined existing FDA tools, resources and guidance to reduce the risk of intentional food contamination. In May 2016, the FSA introduced a further iteration, the Food Defense Plan Builder, a software programme designed to assist food manufacturers to develop personalised food defence plans. This software is available online and includes vulnerability assessment, and the development of a mitigation strategy and associated action plan. It should be noted that a food defence plan is only one aspect of wider controls to mitigate for food crime.

6.6 Decision tools such as vulnerability assessment critical control point (VACCP) and the TACCP approach have been developed based on hazard analysis critical control point (HACCP) principles. There are multiple explanations of how VACCP and TACCP are differentiated from each other in industry texts and standards, and at the time of this new edition of the Guide being written there is no clear focus as to a definitive approach. The aim of this chapter is not to develop a specific critique of this nature, but to explain that many private certification standards, such as the British Retail Consortium (BRC) Global Food Standard (Version 7), require a manufacturer to undertake a vulnerability risk assessment and consider all food raw materials or groups of raw materials at risk of adulteration and substitution (see 5.2 for definitions). This is a much narrower scope than the wider vulnerability risk assessment approaches outlined in this chapter. In this context the BRC Standard (Version 7) describes adulteration as the 'addition of an undeclared material into a food item for economic gain'. The Standard outlines that the following factors need to be taken into account:

- *The nature of the raw material*, e.g. solid, liquid, powder, granular, pure or a mixed ingredient product. These factors will

---

[6] *Vulnerability Assessments of Food Systems. Final Summary Report June 2009 – February 2012* Available from http://www.fda.gov/downloads/Food/FoodDefense/ UCM317547.pdf.

not only affect the potential for adulteration or substitution, but also how such activity could take place, e.g. minced meat ingredients for products such as burgers and ready meals will be at higher risk of substitution than carcase meat as substitution will be less visually obvious when, for example, substituting minced beef with minced pork.

- *Ease of access to raw materials through the supply chain.* The vulnerability of the manufacturing unit and the wider supply chain can be considered using the risk assessment methods described above.

- *Historical evidence of substitution or adulteration.* Some food ingredients have a much higher historical incidence of adulteration, e.g. milk powder, gluten, minced meat, fish and spices. Manufacturers may have limited access to the intelligence surrounding historic evidence of adulteration although some databases do exist regarding adulteration and substitution incidents (see 5.2). Supplier history, in terms of their performance over time, degree of non-compliance with specifications and requirements, reliability in other areas such as legislative compliance with labour laws, accounting practices etc., will give an insight for the manufacturer into supplier behaviour and whether materials derived from that supplier are more likely to be illicit. However, just because a food ingredient does not have a recognised history of adulteration does not mean that the material is necessarily low risk and less vulnerable to illicit activity as the activity may have gone on and simply been undetected.

- *Sophistication of routine testing* to determine food integrity. The degree of sophistication of routine testing needs to be considered by the manufacturer. The manufacturer needs to know what the adulterant is likely to be and an analytical method must be available that will test for that adulterant in that given foodstuff at a sufficient limit of detection that gives validity and confidence to the results. In the fast-evolving world of economically motivated adulteration (EMA) those criteria are often hard to fulfil. Any given testing regime will have a limit of detection and thus it is not possible in many instances to declare that a product is 'free from' a given material. Whilst techniques are being developed to test for certain materials known to be associated with EMA, this technology is in its infancy. When compared to their batch sizes and overall profitability, a small manufacturer may not have the financial ability to access costly analytical tests. From a legal perspective the manufacturer must demonstrate they have precautions in place that are reasonable to ensure that their food complies with labelling and product or process claims (see 1.10). This area is considered more fully in Chapters 15 and 16.

- *Economic factors* may make adulteration or substitution more attractive. There are multiple economic factors that can influence the degree of vulnerability of a given raw material. Supply chain pressure can arise from market dynamics such as the balance between supply and demand, as shown with the melamine incident (where demand for milk powder in China outstripped supply due to resource scarcity), market competition (e.g, where economies of scale force smaller and/or less competitive organisations to consider substitution or adulteration to remain competitive) and financial pressure from diminishing margins or stakeholder pressure to maintain dividends that leads to rationalisation in the minds of those within the organisation that illicit behaviour is somehow acceptable.

6.7   Vulnerability risk assessment and ranking or prioritisation of risks is not a static process. The vulnerability risk assessment needs to be reviewed formally if there is no non-compliance on a minimum annual basis. Reactive review is also essential. Reactive review of the vulnerability risk assessment can arise from intelligence concerning a particular new, emerging or existing threat or a breakdown in security measures in the supply chain, identification of increased vulnerability after verification activities such as an audit that countermeasures and other controls are not effective or the need to purchase raw materials from an alternative source with no supplier history. All reviews should be recorded and any necessary action taken as a result.

6.8   All FCRA documents should form part of the controlled documents system defined by the manufacturing system. These documents should be reviewed at designated intervals to ensure their currency and also in the event of an incident.

7   SECURITY AND COUNTERMEASURES

Principle

Secure buildings and the wider perimeter are key aspects of good manufacturing practice (GMP), especially with regard to crime mitigation. Countermeasure is a term widely used in criminology literature, but only more recently with regard to food science and technology. A countermeasure is quite simply the action taken by an individual, organisation or other body to counteract or offset a given danger or threat. A threat in this context is an agent that can cause harm or loss that arises through the ill-intent of others (see 5.6). A countermeasure is designed to prevent or reduce the impact of a threat through technological, tactical or planned means. Countermeasures can be actions by individuals, or processes or systems that are always operating or can be enacted when a threat is realised.

Background

7.1   Counteracting criminal behaviour in the food supply chain is an embedded requirement of GMP. The use of food crime risk assessment (FCRA) tools to identify potential threats and the means for their control (countermeasures) through the identification of critical points in the manufacturing setting where vulnerabilities need to be eliminated, or if this is not possible effectively managed, has been explored in Chapter 6. The means to then implement appropriate countermeasures and the associated monitoring and verification activities are addressed in this chapter. Ultimately, as an emerging element of GMP, how FCRA is undertaken and countermeasures identified and then integrated into an effective food integrity management system (FIMS) is a crucial area for the food manufacturer to consider and address. At the time of writing this version of the Guide there are a number of International Standards Organisation (ISO) standards being developed under the control of the ISO/TC 292 Security and Resilience Committee. These standards are explained in more detail here and crisis management protocols are addressed in Chapter 27.

ISO 28001 *Security management systems for the supply chain – Best practices for implementing supply chain security, assessments and plans – Requirements and guidance* states there are four types of action a business can take when faced with a threat. These can be called the 4Ts:

- *Treat*: through the use of an organisational procedure and/or physical protection countermeasure;
- *Transfer*: transfer of the risk may be subcontracting of a manufacturing activity, physical transfer of an activity in the manufacturing process to other locations on the site, designated time for activity etc.

Food & Drink – Good Manufacturing Practice: A Guide to its Responsible Management, Seventh Edition.
The Institute of Food Science & Technology Trust Fund.
© 2018 John Wiley & Sons Ltd. Published 2018 by John Wiley & Sons Ltd.

- *Terminate*: due to the level of risk associated with the threat the decision is made not to continue the activity; or
- *Tolerate*: decide to accept the risk associated with the threat because the organisation cannot implement a countermeasure or it is too costly or impractical to implement a countermeasure.

Characteristics of Countermeasures 7.2 Countermeasures can be *global* and operate at the national or supply chain level and counteract single or multiple threats, or they can be *specific* to a particular threat at a particular point in the supply chain or within the manufacturing operation, i.e. they are *situational*. Countermeasures can also be distinguished in terms of whether they address *detection* (i.e. the identification of the activities associated with food crime and potential opportunity), *mapping* of the food chain to identify vulnerabilities or hotspots (e.g. through horizon scanning or other vulnerability risk assessment techniques considering potential threat scenarios associated with the materials, services and final products that the manufacturing organisation procures, produces and sells), *deterrence* (by inhibiting opportunity to commit crime through identifying different kinds of perpetrator and their actions and what countermeasures will discourage their activity; see Chapter 6), *prevention* (through using the resources available to minimise the likelihood and/or severity of a potential threat and promoting the robust systems in place) or *disruption* (should crime occur; see the work of Spink et al.[1]).

Countermeasures can be *passive* or *active* in nature. Passive countermeasures tend to be physical protection or hard resistance in terms of facility design within the manufacturing environment, such as locking mechanisms, self-sealing equipment, enclosed production lines, security systems that involve fencing, access control points for people and vehicles, cameras or other surveillance equipment. This is a term often used when designing food defence systems (see 5.4). Alternatively, with regard to cyber-crime, a firewall is an example of a passive countermeasure. Active countermeasures are reactive, often designed for a particular threat and are enacted once the threat is identified as being 'real' and/or imminent. An example of an active countermeasure or active protection system are those protective actions that are enabled in terms of a cyber threat to a business, such as malware or a hacking attack, or the actions of security personnel in the event of an intruder or unauthorised person entering a controlled area of a factory. Active countermeasures in the manufacturing process itself include intruder alarm systems, access alarms in enclosed production systems and continuous awareness and training programmes and refresher training of individuals so they understand their role in the food manufacturing environment with regard to the range of food crime incidents that could occur.

[1] Spink, J., Moyer, D.C. and Whelan, P. (2016). The role of the public private partnership in food fraud prevention – includes implementing the strategy. *Current Opinion in Food Science*, 10, 68–75.

7.3   *Hurdles* are countermeasures that reduce opportunity for food crime by either assisting detection or proving to be a deterrent.[2] They include online monitoring and verification activities such as audits and product sampling. Examples of hurdles can be differentiated by how they prevent or deter:

- *Access to premises*: Examples include visible and appropriate perimeter fencing and vehicular access points; external landscaping, perimeter alarm system, closed circuit television (CCTV) monitoring/recording of the perimeter at designated vulnerable points or infrared heat monitoring (see 19.2); vehicles parking outside the perimeter, limited access for individuals to the manufacturing site, with zoned and restricted access for individuals within the manufacturing site; monitored vehicular access points; security guards; traffic calming on approach to site, schedule delivery times only; missed deliveries investigated; documentation associated with deliveries checked before vehicular admittance; and monitoring of services access points, e.g. air ventilation and exit points, waste storage and treatment facilities, water systems.
- *Access by people*: Countermeasures include swipe card, password-protected or coded access to site, or specific zones within the manufacturing site for people; protocols for lone working/ buddying system; changing facilities where workwear and personal items are kept separate; screening of visitors – by appointment only, proof of identity required, accompanied visits only; secure handling of mail; restrictions on access for portable electronic and camera equipment, portable hard drives, USB sticks etc.; limited access to mains services; BS ISO/IEC 27032 compliant cybersecurity (see 5.5); employment checks – proof of identity and qualifications, verification protocol for contractors, supervision of temporary staff and contractors, employment checks on those working at designated vulnerable points; staff in critical roles motivated and monitored; whistleblowing arrangements (see 7.10); suitable end of contract arrangements for staff and preventive controls after termination of employment, e.g. changing access, closing computer accounts and access; enclosed conveyors, processing equipment, use of sight-glasses, numbered seals on bulk storage silos and password protection of computer terminals and workstations.
- *Access to product*: Examples include sealing of bulk storage containers, product tampering controls, global positioning systems (GPS) technology, and checks including stock control checks; secure user codes and passwords for access to materials and production lines/zones; electronic controls and reporting in the event of unauthorised access or activity on electronic

---

[2]Spink, J., Moyer, D.C., Park, H., Wu, Y., Fersht, V., Shao, B. and Edelev, D. (2015). Introducing food fraud including translation and interpretation to Russian, Korean, and Chinese languages. *Food Chemistry*, 189(1), 102–107.

systems; supplier assurance protocols and screening protocols for new suppliers, including taking up of references and credit checking. Traceability systems will identify individual lots of raw materials and final product (see Chapter 14).

7.4 *Guardians* are the individuals operating at national, supply chain or individual business levels that have the knowledge, skills and understanding to implement a FIMS. They may operate within the manufacturing environment or in the wider supply chain. It is important to recognise that vulnerability can still occur even in the presence of a capable guardian (see Spink et al., 2015[2]).

New Product and New Process Development

7.5 Products are more vulnerable when specific claims are made with regard to the product, e.g. designated provenance, identity or assurance status (e.g. Fairtrade, organic, country of origin, protected geographical indication (PGI); see Chapter 15). When designing new products consideration should be given during the product development phase as to their potential vulnerability to food fraud or food crime (see 5.1). If required, specific design objectives with regard to food integrity should be developed. When the packaging or product is designed, inherent 'markers' could be included/adopted so that the product can be readily identified as being authentic and not counterfeit or substituted. During the development stage, developing a unique spectral fingerprint for the product should also be considered so that its integrity can be verified. With this approach the exact nature of the substitute or adulterant does not need to be known, only that product integrity has been lost.

7.6 Process development activities should consider the potential for 'engineering out' parts of the process where tampering or substitution could occur, e.g. by incorporating hard resistance countermeasures such as the use of machine covers, guards or enclosed sections.

Security Management Systems

7.7 ISO 28000:2007 *Security management systems for the supply chain* specifies requirements for a security management system, including those aspects critical to security assurance of the supply chain, including transport and logistics. ISO 28000:2007 requires the manufacturing organisation to assess the security environment in which it operates and to determine if adequate security measures are in place. This standard, along with many other ISO standards, is based on an approach known as plan-do-check-act (PDCA). The four steps are:

Plan: Establish the objectives and processes necessary to deliver results in accordance with the organisation's security policy.

Do: Implement the objectives and processes.

Check: Monitor and measure processes against security policy, objectives, targets, legal and other requirements, and report results.

Act: Take corrective or preventive actions to continually improve performance of the security management system.

ISO 28001 defines a security assessment as identifying and documenting supply chain vulnerabilities to a set of defined security threat scenarios as well as the likely persons who could progress a particular security threat into an actual incident. A security plan outlines the security measures (countermeasures) that are in place, or need to be introduced, to manage the security threat scenarios identified in the security assessment and reduce the threats identified in the security plan to an acceptable level. A training programme needs to be formally developed and implemented that identifies how security personnel will be trained according to their work role.

There are five distinct elements to a security management system (SMS): the security management policy (and associated security objectives), the security assessment (documented assessment that identifies the threats considered to be of importance), implementation and operation of the SMS, monitoring, verification and checking of the SMS, and management review and continuous improvement.

Annex B of ISO 28001 outlines the risk management methodology in eight steps:

1. Identify all activities within the scope of the SMS.

2. Identify the security controls and countermeasures currently in place.

3. Identify security threat scenarios.

4. Determine the consequences if the security threat scenario actually occurred.

5. Determine the likelihood of such an event occurring, given the security controls and countermeasures currently in place.

6. Assess whether the current security controls and countermeasures are adequate.

7. If they are not adequate, develop and implement additional controls and countermeasures (develop a security plan).

8. Repeat the process at a defined interval, e.g. 6-monthly or annually.

7.8   A documented security assessment should be undertaken on the manufacturing site to determine the security procedures required in view of the site layout, nature of the products produced and the processing employed, and the specific area of the site. It may be prudent to identify high-risk (HR) and low-risk (LR) areas with regard to product security and have additional security controls in place in HR areas. This can include, but is not limited to, colour of protective clothing for HR areas, restricted entry procedures for authorised personnel only, including fingerprint entry to HR areas, and site security arrangements as a whole in terms of fencing, security staff and close-circuit television (CCTV). Keys that are required for entry to HR areas should be subject to a sign out and sign back in control protocol.

The possibility of sabotage, vandalism, terrorism and even site invasion may indicate a need for particular security precautions in vulnerable areas, e.g. building entrance security, code pads to open external doors to manufacturing areas, locked rooms, use of seals, etc. Access to material and product storage areas should be restricted to personnel working in those areas and other authorised persons. Consideration should be given to site security and entry controls into designated storage areas. Designated separate lockers may need to be provided for both internal and external workwear as well as personal belongings that must not be taken into the production and storage areas.

A visitor and contractor reporting system should be in place and the person responsible for such individuals is also responsible for monitoring their activities when they are on site. The use of photographic and recording equipment, including mobile equipment that contains these functions, must be strictly controlled. All such equipment brought onto the production site must be authorised by the site management, e.g. memory sticks, laptops, tablets, mobile phones etc. Staff should be encouraged to report unknown individuals when they see them on site, but be made aware, through induction and refresher training, of the personal dangers of challenging unknown individuals. The security assessment should be reviewed a minimum of annually and more often in the event of a breach of security arrangements or identification of emerging threats or a change in threat status. This security assessment can then be used when considering wider organisational vulnerability as part of overall FCRA (see Chapter 6).

7.9   Appropriate training must be given to all staff on the actions to take in the event of them uncovering criminal activity taking place or highlighting the evidence of past criminal activity (see 7.10). The personal safety of staff is paramount. There is a duty of care on senior management to ensure appropriate procedures are in place to safeguard staff who bring forward concerns or evidence of food crime having taken place. This will include working with the relevant regulatory authorities to support such individuals. Appropriate procedures also need to be in place for those staff members who are vulnerable, e.g. those who are working alone or undertaking remote working activities at the processing site, and/ or those working with high-value food materials and products. There is also the potential for perpetrators to be embedded in an otherwise legitimate business. The category of individual that perpetrates a crime will influence the countermeasures that need to be in place to mitigate against their activity (see 5.7).

Clear instructions must be given during training on what activities are legal within the food manufacturing environment and what would constitute illegal activity where individuals as well as the food business could be prosecuted. The training programmes must be tailored for the job role and the particular vulnerabilities of interest, e.g. supplier approval and performance monitoring, procurement, operational management etc. Refresher training must be undertaken if intelligence suggests that there are new potential risks that staff need to be aware of, e.g. adulterated ingredients, counterfeit packaging etc. Training should be reinforced by adequate supervision, mentoring and support, and regular performance reviews. Training is addressed more fully in Chapter 17. If deemed necessary in security assessment, fingerprint technology should be used so that individuals are traceable to specific production areas and, by association, specific batches of product.

Consideration should be given to organisational incentive bonus schemes which could inadvertently promote occupational food crime. If implemented, such schemes should be designed to discourage operators from taking unauthorised shortcuts and the activities monitored through effective supervision. The potential for inappropriate action by current or previous employees needs to be addressed by the manufacturer including pre-employment screening and employment termination interviews.

Whistleblowing   7.10   Personnel should be instructed and encouraged to report immediately any incident or potential incident associated with people, materials, packaging, equipment or the product. They should be aware of the potential for illegal behaviour at all stages of the manufacturing operation and be in a supportive culture and environment where they feel confident to report any issues that may arise.

In simple terms, a whistleblower is any person who reports or discloses information about a given threat or of harm to the public interest that is identified in the context of their work (Council of Europe, 2014).[3] Publicly Available Specification (PAS) 1998 (2008:9) *Whistleblowing arrangements: Code of practice*[4] defines a whistleblowing concern as a 'reasonable and honest suspicion an employee has about a possible fraud, danger [illegality] or other serious risk that threatens customers, colleagues, shareholders, the public or the organisation's own reputation'. This is different to an employee grievance or complaint, which PAS 1998:2008 states is an employment dispute that has no public interest element.

---

[3]Council of Europe (2014). *Recommendation CM/Rec (2014) 7 of the Committee of Ministers to member States on the protection of whistleblowers.* Available at https://rm.coe.int/1680700282.
[4]https://shop.bsigroup.com/forms/PASs/PAS-1998/.

The act of whistleblowing can be *internal*, i.e. disclosure within the organisation to a supervisor or manager, or *external* to regulatory officers, the police, an independent body, media, auditor or consumer group and so forth. The cost to the organisation of a missed opportunity to treat, transfer or terminate a threat can be significant in terms of loss of business, profits and jobs, fines and an increase in insurance premiums.[5]

If the food manufacturer wants to effectively counteract illegal practice within the organisation and wider supply chain, then organisational whistleblowing procedures should be adopted. Internal reporting channels should be available for staff so that action can be taken to prevent or reduce the likelihood of criminal activity and minimise the impact on customers and consumers. Adopting procedures that identify the communication channels for whistleblowing to take place and also to protect individuals should they whistleblow will strengthen this approach as a countermeasure.

The European Committee on Legal Co-operating (CDCJ) of the Council of Europe developed the Recommendation CM/Rec (2014)7 on the protection of whistleblowers (Council of Europe, 2014). Member states are encouraged to develop a robust national framework that facilitates and protects whistleblowers in all industries. The Recommendation sets out a number of key principles to ensure that at a regulatory level there are:

- laws to protect whistleblowers cover a broad range of information that is in the public interest;
- access points to more than one communication channel for individuals to report and disclose sensitive information;
- mechanisms in place to ensure reports and disclosures are acted upon promptly; and
- protocols to ensure whistleblowers' identities remain confidential and all forms of retaliation are prohibited as long as the individual whistleblower has reasonable grounds to believe in the accuracy and credibility of the information.

In the UK, the **Public Interest Disclosure Act 1998**, and equivalent legislation, protects workers, if they report wrongdoing in the workplace, from unfair treatment or victimisation from their employer. If they raise concerns in accordance with the Act's provisions, employees are protected under the Act.[6] The Food Standards Agency's National Food Crime Unit (NFCU) was created as a result of a recommendation in the 2014 Elliott

---

[5]Soon, J.M. and Manning, L. (2017). Whistleblowing as a countermeasure strategy against food crime. *British Food Journal*, 119(12), 1–25.
[6]UK Food Standards Agency. *Reporting food fraud*. Available at https://www.food.gov.uk/enforcement/the-national-food-crime-unit/foodfraud.

Review.[7] All intelligence received by the NFCU is logged on the Food Fraud Database and an investigation launched.

The UK Department of Business, Innovation and Skills report *Whistleblowing: Guidance for Employers and Code of Practice*,[8] issued in 2015, states it is good business practice to implement a whistleblowing policy. PAS 1998:2008 outlines that overall responsibility for whistleblowing should rest with the Board, Chief Executive, Group Secretary, legal department or finance department, with day-to-day responsibility falling to the human resources department, who can implement a formal protocol. In smaller manufacturing businesses a simple statement will be sufficient. The statement should explain what whistleblowing is and that it should not be a means for undermining managers, how to make an internal or external disclosure, e.g. on an independent disclosure helpline, and the process for maintaining confidentiality. The statement or larger policy should be communicated to staff so that they are aware of the requirements. For some manufacturers their customers may have prescribed processes for whistleblowing. This may include awareness training, the siting of posters and information data, and internal reviews of staff engagement with the protocols.

[7] Elliot Review (2014). *Elliot review into the integrity and assurance of food supply networks – final report. A national food crime prevention framework*. Available at https://www.gov.uk/government/uploads/system/uploads/attachment_data/file/350726/elliot-review-final-report-july2014.pdf.
[8] UK Department of Business, Innovation and Skills (2015). *Whistleblowing: Guidance for Employers and Code of Practice*. UK Department of Business, Innovation and Skills Report, London.

Principle

As part of their natural protection mechanism to survive adverse growing conditions, insect attack or other kinds of hostile environment, plants or fungi can naturally produce a variety of toxic chemicals. These can be toxic when consumed by humans, and in some cases be mutagenic or carcinogenic. The manufacturer needs to be aware of the potential risk of plant toxins and have appropriate management systems in place to minimise the risk to consumers of the food products they produce. Great care should be taken (a) to formulate foods to avoid, wherever possible, inclusion of major serious allergens as ingredients, (b) to provide appropriate warning to potential purchasers of the presence of a major serious allergen in a product and (c) to organise production, production schedules and cleaning procedures to prevent cross-contamination of products by 'foreign' allergens.

Plant Toxins

8.1   Naturally occurring plant-related toxins include oxalic acid and prussic acid. Those produced by fungi and moulds include alternaria, ergot and mycotoxins. Algal toxins, which can be concentrated in seafood, include okadaic acid and domoic acid. As an example, solanine and chaconine are glycoalkaloids naturally produced by potatoes as a defence mechanism against insects, pathogens and herbivores on potato tubers exposed to light. The associated 'greening' of the tuber, caused by chlorophyll production, suggests increased levels of glycoalkaloid within the potato, meaning such potato tubers are not fit for human consumption. At the product development stage all plant-derived ingredients must be considered for the potential of these naturally occurring toxins to cause acute or chronic harm to consumers and appropriate controls must be in place to minimise harm to the consumer. These controls may need to cover growing, harvest, storage, transport or processing conditions.

The examples used in this chapter are not designed to be exhaustive, but give an insight into the manufacturer's obligation to understand the potential for natural toxins to occur in the raw ingredients or products they handle and process, and confirm that the onus is on the manufacturer to ensure that appropriate controls are in place to safeguard the consumer from harm. The manufacturer must have appropriate food safety risk assessment processes in place, e.g. hazard analysis critical control point (HACCP; see Chapter 3), that consider these natural toxins if they are likely to occur in ingredients or products that are to be processed and how the risk of harm to consumers can be mitigated. Raw beans, for example, contain lectin, a toxic agent that is inactivated by appropriate processing conditions. Manufacturing processes for products that contain beans therefore must be developed to inhibit lectin activity to ensure the food is then safe to consume.

Food & Drink – Good Manufacturing Practice: A Guide to its Responsible Management, Seventh Edition.
The Institute of Food Science & Technology Trust Fund.
© 2018 John Wiley & Sons Ltd. Published 2018 by John Wiley & Sons Ltd.

8.2 Mutagens are physical or chemical agents that change the genetic material of an individual and in so doing increase the number of genetic mutations that occur in its body. Naturally occurring mutagens are ubiquitous in processed foods, particularly cooked and smoked foods. It is the responsibility of a manufacturer to be aware of the potential presence of these mutagens and to ensure that appropriate procurement procedures are in place to minimise the risk to consumers. As many mutagens can cause mutations that can lead to cancer, these particular agents are also classified as carcinogens. Again this topic is explored through an example rather than giving an exhaustive account within this chapter. Acrylamide is an example of a carginogenic mutagen that can be found in some fried, roasted or baked carbohydrate-based foods, or more generally foods heated to above 120 °C. Manufacturers who undertake these process steps with the carbohydrate-based foods they manufacture should undertake a specific acrylamide risk assessment in order to develop risk-based controls that are effectively implemented and routinely verified to ensure their efficacy. A minority of mutagenic processing by-products have legal limits, but in all cases the underlying principle of risk management is an ALARA (As Low As Reasonably Achievable) approach.

The senior management of the manufacturing organisation should ensure that adequate resources are committed to the development of a comprehensive food safety management system (FSMS) that is designed, documented, implemented and reviewed and so furnished with personnel, equipment and resources to ensure that all potential food safety hazards are adequately controlled and mitigated (see Chapter 3).

Adverse Reactions to Foods

8.3 More widespread adverse reactions to foods and food ingredients not only include allergic reactions, but also reactions such as functional food intolerance due to enzymatic abnormalities (enzymopathy) in individuals, e.g. lactase deficiency leading to lactose intolerance, structural intolerance affecting the gastrointestinal tract (e.g. transport defects in the gut, fructose intolerance) and pharmacological reactions to non-proteinaceous substances (e.g. colourants (tartrazine), preservatives (benzoates or sorbates)), sodium glutamate or amines due to excessive intake from food rich in tyramine, tryptamine, histamine and serotonin. Histamine, for example, is said to play a role in migraines in susceptible individuals. Histamine-rich and histamine-releasing foods include fermented foods, processed meats, wine, beer, cheese, chocolate and dried fruits such as apricots.

8.4 The United Kingdom (UK) Food Standards Agency (FSA) has highlighted six food colours that have been associated with intolerance and hyperactivity in young children. These are sunset yellow FCF (E110), quinoline yellow (E104), carmoisine (E122), allura red (E129), tartrazine (E102) and ponceau 4R (E124). A voluntary ban has been introduced, and manufacturers have been asked to reformulate products to remove these colourants.

In respect of these six colours, the **EU Regulation No. 1333/2008**[1] on food additives requires that, as of 20 July 2010, for all foods that contain these colours a warning is included on the packaging to the effect that the colours may have an adverse effect on activity and attention in children. The FSA provide further guidance on the labelling associated with these colours.[2] **Annex II on EU Regulation No. 1338/2008** was published as **Commission Regulation (EU) No. 1129/2011**, entered into force on 2 December 2011 and applied from 1 June 2013. **Regulation (EU) No. 232/2012**, which came into force on 23 May 2012 and applied from 1 June 2013, amends Annex II to restrict the use and levels for three colours: quinoline yellow, sunset yellow and ponceau 4R. The levels of these colours are now restricted in a number of food categories, including soft drinks, confectionery, sauces and seasonings and in some cases (e.g. ponceau 4R in sauces and seasonings) no longer permitted. The Regulation includes a maximum use level of 20 mg/l for sunset yellow in soft drinks. Exemptions with regard to 'warning' labelling are in force in the UK, e.g. for alcoholic drinks above an alcohol content of 1.2%.

8.5   About 1–2% of adults and between 5 and 7% of children are thought to suffer from some type of immunoglobulin E (IgE)-mediated food allergy. The higher incidence of allergies in infants is due to allergy to cow's milk, which children generally grow out of by school age.[3] Individuals can experience both immunological and non-immunological adverse reactions to food. Adverse reactions to food that have an immunological basis are termed 'food allergies' and include those that involve IgE-mediated reactions and the auto-immune conditions requiring gluten avoidance, coeliac disease, which is thought to have a cellular immune mechanism. This means that the problems associated with food allergens are part of a wider problem that includes all adverse reactions to foods. These include reactions that result from microbial and chemical food poisoning, psychological aversions, behaviour change and specific non-allergenic responses. It is important for the manufacturer to understand that these immune-mediated (allergic) and non-immune-mediated reactions exist and that they need to demonstrate that they have done all that could reasonably be expected to consider the health impacts of the food products they manufacture on their consumers.

In total over 170 foods have been documented in the scientific literature as causing allergic reactions. However, most plant allergens belong to a small number of protein superfamilies. Food allergy is a complex topic and multiple factors influence which individuals may exhibit an allergic reaction to a given food. Scientific papers report that these include nationality, dietary

[1] http://eur-lex.europa.eu/legal-content/EN/TXT/PDF/?uri=CELEX:02008R1333-20160525&qid=1480604 927473&from=EN.
[2] https://www.food.gov.uk/sites/default/files/multimedia/pdfs/labellingcoloursreg13332008.pdf.
[3] http://www.ifst.org/knowledge-centre/information-statements/food-allergy.

habits and hereditary or environmental factors.[4] As food can be sold across the world it is a requirement for food manufacturers to understand in detail the issues that can arise in a particular export market with regard to labelling of foods or food ingredients that have been identified as being associated with an allergic response in sensitive individuals. Cross-reactivity may occur where an individual is allergic to a protein in one food, which resembles a protein found in something else e.g. individuals sensitised to latex (rubber) can have a similar response to some proteins present in kiwi fruit, banana or avocado, whilst those sensitised to peanut many have an allergic response to lupin flour.

Proteins that cause an adverse response in susceptible individuals can be divided into groups or families. The *prolamin* superfamily of proteins, produced as a response to an insect attack, is associated with cereals and the *cupin* superfamily is associated with tree nuts, seeds and legumes such as chickpeas. The *profilin* family (e.g. melon) and *pathogenesis-related* (PR) proteins are not families, but are a group of proteins produced as part of the plant defence mechanism. These latter proteins can be found, for example, in celery and stone fruits. There is a differentiation within the law for the supply chain controls that are required. As outlined there are foods, ingredients or proteins known to cause an adverse reaction in susceptible individuals. More specifically there are food ingredients and foods themselves known to cause an allergic response in susceptible individuals, and these foods are required to carry allergen labelling by law. The specific legislative requirements of which foods require allergen labelling and which do not differ from country to country.

8.6   Manufacturers need to make themselves more aware of the wider issues surrounding allergenic response to food and more generally adverse reactions to food and food ingredients to ensure that they have followed reasonable and appropriate protocols to meet legislative requirements and also the wider social interest in food allergy and intolerance. More detailed information on allergenic foods can be found in the InformAll database (http://research. bmh.manchester.ac.uk/informAll).

Effective management of serious food allergens is an essential part of good manufacturing practice (GMP). This is an extremely complex problem to which there are no cheap or easy solutions. As outlined, there are few foods or food ingredients to which someone, somewhere, is not allergic or intolerant, in some cases in very small (microgram) quantities. Allergic reactions may range from relatively short-lived discomfort to anaphylactic shock and death, but all should be treated seriously and safeguarded against by the manufacturer. While not detracting from

---

[4] Manning, L. and Soon, J.M. (2017). An alternative allergen risk management approach. *Critical Reviews in Food Science and Nutrition*, 57(18), 3873–3886.

the responsibility of sufferers (and their medical advisers) to identify the particular foods or food substances to which they are allergic, there is a need for due diligence by manufacturers, firstly in considering the use of known allergens as ingredients, then in warning the customer or consumer of the presence or potential presence of such allergens, and thirdly in preventing accidental cross-contamination of products by allergens used in other products. This is not only a duty of care and a due diligence requirement, but an essential means of minimising the risk of being subject to a product liability claim. While the guidelines outlined in this chapter should prove useful in providing essential signposts towards developing organisational GMP in this area, the new development of such a policy requires commitment in the company 'culture', an allocation of funds and resources, and a concentrated and sustained effort by everyone led by senior management, and its application and maintenance thereafter.

8.7   Appropriate warnings to the potential purchaser are necessary, which involves labelling. The use of labelling should not be regarded as obviating the responsibilities of sufferers (and their medical advisers) to identify the particular foods or food substances to which they are allergic, or the responsibilities of manufacturers referred to in the following paragraphs.

Legislative 8.8   The Codex Alimentarius Commission Committee on Food
Requirements   Labelling has listed the foods and ingredients that cause the most severe reactions and most cases of food hypersensitivity. Section 4.2.1.4 of General Standards for the Labelling of Prepackaged Foods states that 'the following foods and ingredients… shall always be declared: cereals containing gluten; i.e., wheat, rye, barley, oats, spelt or their hybridized strains and products of these; crustacea and products of these; eggs and egg products; fish and fish products; peanuts, soybeans and products of these; milk and milk products (lactose included); tree nuts and nut products; and sulphite in concentrations of 10 mg/kg or more'.[5]

In the European Union, **EU Regulation (EU) No. 1169/2011** on the provision of food information to consumers has been mandatory in all Member States since 13 December 2014.[6] It was made mandatory on 13 December 2016. Annex II list the allergens that must be declared and some exemptions, as follows:

1.   cereals containing gluten, namely, wheat, rye, barley, oats, spelt, kamut or their hybridised strains, and products thereof, except:

[5]Codex Alimentarius Commission (1985). *General Standard for the Labelling of Prepackaged Foods (CODEX STAN 1-1985)*. Available at http://www.codexalimentarius.org.
[6]http://eur-lex.europa.eu/LexUriServ/LexUriServ.do?uri=OJ:L:2011:304:0018:0063:EN:PDF.

(a) wheat-based glucose syrups, including dextrose,[7]
(b) wheat-based maltodextrins,
(c) glucose syrups based on barley,
(d) cereals used for making alcoholic distillates, including ethyl alcohol of agricultural origin;

2. crustaceans and products thereof;
3. eggs and products thereof;
4. fish and products thereof, except:
   (a) fish gelatine used as carrier for vitamin or carotenoid preparations,
   (b) fish gelatine or Isinglass used as fining agent in beer and wine;
5. peanuts and products thereof;
6. soybeans and products thereof, except:
   (a) fully refined soybean oil and fat,
   (b) natural mixed tocopherols (E306), natural D-alpha tocopherol, natural alpha tocopherol acetate and natural D-alpha tocopherol succinate from soybean sources,
   (c) vegetable oils derived from phytosterols and phytosterol esters from soybean sources,
   (d) plant stanol ester produced from vegetable oil sterols from soybean sources;
7. milk and products thereof (including lactose), except:
   (a) whey used for making alcoholic distillates, including ethyl alcohol of agricultural origin,
   (b) lactitol;
8. nuts, namely, almonds (*Amygdalus communis* L.), hazelnuts (*Corylus avellana*), walnuts (*Juglans regia*), cashews (*Anacardium occidentale*), pecan nuts (*Carya illinoinensis* (Wangenh.) K. Koch), Brazil nuts (*Bertholletia excelsa*), pistachio nuts (*Pistacia vera*), macadamia or Queensland nuts (*Macadamia ternifolia*) and products thereof, except for nuts used for making alcoholic distillates, including ethyl alcohol of agricultural origin;
9. celery and products thereof;
10. mustard and products thereof;
11. sesame seeds and products thereof;
12. sulphur dioxide and sulphites at concentrations of more than 10 mg/kg or 10 mg/L in terms of the total SO_2, which are to be calculated for products as proposed ready for consumption or as reconstituted according to the instructions of the manufacturers;
13. lupin and products thereof; and
14. molluscs and products thereof.

Commission Delegated Regulation (EU) No. 78/2014 amends **Annex II of the EU Food information to Consumers** listing the 14 allergens. This list is subject to future amendment. It does not preclude the manufacturer from drawing attention on the label to

---

[7]And the products thereof, in so far as the process that they have undergone is not likely to increase the level of allergenicity assessed by the authority for the relevant product from which they originated.

other food allergens. **Article 21 of EU 1169/2011 on labelling of certain substances or products causing allergies or intolerances**, states that:

1. Without prejudice to the rules adopted under Article 44(2), the particulars referred to in point (c) of Article 9(1) shall meet the following requirements:
    (a) they shall be indicated in the list of ingredients in accordance with the rules laid down in Article 18(1), with a clear reference to the name of the substance or product as listed in Annex II; and
    (b) the name of the substance or product as listed in Annex II shall be emphasised through a typeset that clearly distinguishes it from the rest of the list of ingredients, for example by means of the font, style or background colour. In the absence of a list of ingredients, the indication of the particulars referred to in point (c) of Article 9(1) shall comprise the word 'contains' followed by the name of the substance or product as listed in Annex II. Where several ingredients or processing aids of a food originate from a single substance or product listed in Annex II, the labelling shall make it clear for each ingredient or processing aid concerned. The indication of the particulars referred to in point (c) of Article 9(1) shall not be required in cases where the name of the food clearly refers to the substance or product concerned.
2. In order to ensure better information for consumers and to take account of the most recent scientific progress and technical knowledge, the Commission shall systematically re-examine and, where necessary, update the list in Annex II by means of delegated acts, in accordance with Article 51.

8.9   In April 2015, the UK's FSA issued the *Food allergen labelling and information requirements* under the EU Food Information for Consumers Regulation No. 1169/2011: Technical Guidance that cover the interpretation and application of allergen provisions for prepacked foods, prepacked foods for direct sale and non-prepacked foods as outlined in the EU Food Information for Consumers Regulation (No. 1169/2011).[8] This guide provides guidance to food businesses on allergen labelling.

8.10 In the United States of America (USA), regulation is by the Food Allergen Labelling and Consumer Protection Act of 2004 (FALCPA). From January 1 2006, all packaged foods regulated under the Act must comply with its food allergen labelling requirements if they include milk, egg, fish, crustacean shellfish, tree nuts, wheat, peanuts, soybeans (or protein derived from any of them). The US Food and Drug Administration (FDA) has produced an online *Guidance for Industry: Questions and Answers*

[8] https://www.food.gov.uk/sites/default/files/food-allergen-labelling-technical-guidance.pdf.

Regarding Food Allergens, including the **Food Allergen Labeling and Consumer Protection Act of 2004**[9].

8.11   Exotic and novel foods are also liable to allergen labelling, e.g. products containing insects and/or their derived protein where this is known to cause an allergic reaction in individuals who are susceptible.

8.12   Existing or proposed new product formulations should be carefully examined at the product development stage to see whether there is a possibility of excluding food allergens. It is important to keep up to date with information from consumer organisations and clinicians to be aware of new food allergies and those increasing in prevalence within the population. Of course, in many cases a food allergen is essential to characterise the food, and in such cases label warning must suffice. In some cases, however, where food allergens are present as non-characterising or minor ingredients, it may be possible to effect a substitution.

8.13   Exporters should be aware that legislative requirements for allergen labelling differ from country to country and that pre-packed products must comply with the designated labelling in the country in which the food is sold.

8.14   Inclusion of the name of a food allergen in an ingredients list should not be regarded as adequate warning and current legislative requirements for highlighting allergens must be followed. There are minimum font size requirements for allergen such labelling.[10]

8.15   Where a product nominally free from food allergen is produced on a production line shared with a food allergen-containing product, a suitable warning might be, for example, '**May contain traces of PEANUT**'. However, the use of 'may contain' should not be used as a way of avoiding the measures set out in Sections 8.18–8.25.

8.16   Where a product nominally free from food allergens is produced in the same factory building as a food allergen-containing product, a suitable warning might be, for example, '**Produced in a factory where PEANUT is also handled**'. Again, this should not be used as a way of avoiding the measures set out in 8.18–8.25.

8.17   Some products may be specifically identified as 'allergen-free' products, such as 'gluten free'. **EC Regulation No. 828/2014** on

[9] https://www.fda.gov/food/guidanceregulation/guidancedocumentsregulatoryinformation/allergens/ucm059116.htm.
[10] https://www.food.gov.uk/sites/default/files/food-allergen-labelling-technical-guidance.pdf.

the requirements for the provision of information to consumers on the absence or reduced presence of gluten in food outlines the requirements for the provision of information to consumers on the absence or reduced presence of gluten in food The Regulation states 'the statement "gluten-free" may only be made where the food as sold to the final consumer contains no more than 20 mg/kg of gluten'. The claim of 'very low gluten' can only be made where the food contains no more than 100 kg/mg of gluten in the food as sold to the final consumer. and which contain cereal ingredients that have been specially processed to reduce the level of gluten.[11]

Cross-contact 8.18 The FDA *Approaches to Establish Thresholds for Major Food Allergens and for Gluten in Food*,[12] produced in 2006, describes cross-contact as occurring when 'as part of the food production process where residues of an allergenic food are present in the manufacturing environment and are unintentionally incorporated into a food that is not intended to contain the food allergen, and thus, the allergen is not declared as an ingredient on the food's label'. The accidental presence of a food allergen in a product may arise in three main ways: accidental misformulation, poor labelling/packaging control leading to the product containing an allergen being packed or labelled incorrectly, or cross-contamination by a food allergen from a different product. Misformulation resulting in the inclusion of a food allergen (or any other ingredient) not in the product formulation should be prevented by proper attention to the provisions of Chapter 10 to ensure that the products as prepared contain only the ingredients specified. Labelling control procedures are addressed in Chapter 37.

8.19 Cross-contamination of a product by a food allergen via cross-contact from a different product may arise as a result of a variety of activities but could be due to residues in shared equipment, airborne dust or the improper incorporation of rework material without consideration of the allergen problem. The particulate nature of the allergen should also be considered in any allergen risk assessment as this will affect the potential for contamination and the area over which contamination could occur, i.e. is the allergen, or allergen-carrying food, a liquid, powder, solid layer or granule? It should also be considered whether the allergen is of a homogenous distribution through the food or heterogeneous as this will affect the accuracy and repeatability of sampling and monitoring activities. The allergen risk assessment must integrate with the HACCP plan and the FSMS in its entirety. This will lead to the adoption of appropriate control measures for the manufacturing situation, including a specific allergen control operational pre-requisite programme (PRP).

[11] https://www.food.gov.uk/sites/default/files/multimedia/pdfs/glutenguidance2012.pdf.
[12] http://www.fda.gov/downloads/Food/IngredientsPackagingLabeling/UCM192048.pdf.

8.20   In companies producing on more than one site, or in different buildings on the same site, serious consideration should be given to production segregation in separate buildings. Where separate buildings are not available, separate production equipment or timing separation is recommended. Where production equipment is shared between one or more food allergen-free products and a food allergen-containing product and this is unavoidable, the food allergen-containing product should be run as the last production of the day, immediately before cleaning (e.g. on a shared production line for mixed breakfast cereals, one of which contains nuts, the product containing nuts should be run last). However, it should be recognised that cleaning afterwards, especially in a plant producing dry products, will not necessarily guarantee against small quantities of trapped material waiting to be 'carried over' into the first product to go through, and segregation may be the only acceptable solution. The same applies to small quantities of food allergen in airborne dust. In some instances, air-handling systems might need to be considered.

8.21   As an example of the measures outlined in 8.20, a formal documented allergen risk assessment should be implemented on sites where peanuts or nuts, or any other allergen in fact, are processed or stored to ensure that products containing them do not contaminate peanut-free or nut-free product. Nuts should be stored in a designated area and processed on designated lines. Production and cleaning scheduling (time separation) should be considered between nut-free and nut products if they are processed on the same line. The required cleaning procedures, undertaken after processing nut products to minimise risk of cross-contact, should be validated and implemented. On an ongoing basis clean-down procedures should be verified for efficacy, for example by means of swab tests. A potential source of contamination could be a spillage of nut products onto the packaging of nut-free products. This contaminated packaging might then cause contamination of nut-free products. If a spillage of nuts occurs, this must be cleared up with care, including the use of designated cleaning equipment and disposable wear, or other suitable means, for personnel to ensure that their protective clothing is not contaminated, which would then present a risk to other products. Personnel must ensure that all areas surrounding the spillage are checked for nut debris. All nut debris must be taken immediately to the appropriate disposal area. Any affected goods should be quarantined and disposed of. While nuts have been used in this example, the allergen control procedure should address all likely allergens or allergenic material that could be present in the food manufacturing premises.

The allergen control procedures of suppliers should also be considered so that the likelihood of contamination of raw materials, ingredients or packaging is minimised. Allergenic materials may also arise from processing aids, maintenance and repair activities and factory product trials, and controls should be in place to address all these potential sources. The allergen control risk assessment must also consider the likelihood of contamination and therefore the risk of staff, visitors or contractors bringing allergen-containing food on-site for their own consumption and/or the potential impact of catering functions on-site. Measures should also be put in place to alert staff and site visitors to the presence of allergens on site and to prevent unnecessary exposure to those at risk.

8.22 The incorporation of rework material in a product is covered in Chapter 29, and its provisions should be operated in order to exclude from any food product not containing certain allergens, rework material that contains any of those allergens.

8.23 Formal allergen control training procedures should be in place to train all personnel so they adopt the necessary measures and understand the reasons for why they need to do so.

Quantifying Allergens in Foods 8.24 The Institute of Food Science and Technology (IFST) *Information Statement on Food Allergy* highlights that in order to develop the most cost-effective food allergen management systems it is important to quantify the amount of a food allergen needed to trigger an adverse reaction in the body, which allergen(s) in the food must be monitored and how the allergen behaves during processing.[13]

8.25 In New Zealand and Australia, the membership-based Allergen Bureau was established in 2005. The Allergen Bureau's Voluntary Incidental Trace Allergen Labelling (VITAL) Programme[14] is a standardised allergen risk assessment tool that can assist an organisation to produce an allergen assessment based on a given manufacturing situation and produce a *labelling outcome* that not only addresses intentional inclusion, but also the potential for cross-contact in a resultant precautionary statement: '**May be present: XXX**'.[15] Analytical methods are being developed for allergen quantification but their cost currently prohibits their use as a monitoring procedure, although they form a useful verification tool. These methods include polymerase chain reaction (PCR) assays, enzyme-linked immuno-sorbent assays (ELISA) and mass spectrometry.

As there are no economic analytical techniques to positively release food on a batch-by-batch basis, preventive measures are critical to address allergen control from the product development stage through to product labelling on consumer purchase. These measures need to be reviewed in the instance of non-conformance being identified so that they continue to be effective.

---

[13] http://www.ifst.org/knowledge-centre/information-statements/food-allergy.
[14] http://allergenbureau.net/wp-content/uploads/2013/11/VITAL-Guidance-document-15-May-2012.pdf.
[15] http://allergenbureau.net/vital/.

9   FOREIGN BODY CONTROLS

Principle

The protection of food against contamination with foreign bodies requires the use of hazard analysis critical control point (HACCP) to identify potential sources, with assessment of the types of foreign bodies associated with them and their degree of seriousness. It is important to determine if the foreign bodies are intrinsic, that is, derived from the product, for example fruit stones or fish bone, or extrinsic, that is, derived from the manufacturing environment, as the method of control will be different. Preventive methods are progressively applied at various points in the process flow, manufacturing, packaging, storage and distribution chain to minimise the risk of the presence of foreign bodies in the product. While the use of automatic inspection devices (metal detectors, X-ray machines and vision systems) is recommended as appropriate, it must be remembered that none of these devices is capable of detecting all foreign body contaminants. The major emphasis must always be prevention. Foreign body control procedures are a key prerequisite to ensuring good manufacturing practice (GMP).

Sources of Foreign Bodies

9.1   It is convenient for practical control to divide sources of foreign bodies into those external to the manufacturing plant and those within the plant and premises. Incoming materials and their packaging originating from external sources then immediately become potential internal sources as they enter the manufacturing premises. The types of physical contamination that need to be considered in any food safety management system (FSMS) include glass, ceramic, plastic (hard and soft), wood, metal, paint, paper, cardboard, string, stones, pests and parts of pests, building materials and human-origin foreign bodies.

9.2   External sources are frequently associated with characteristic contaminants such as pest predators on fruits and vegetables or parasites in animals. Similarly, particular methods of production, handling and packaging of incoming materials can give rise to characteristic foreign bodies, for example metal or plastic tags in carcass meat, stones in root crops or slivers of wood in herbs or tea packed in wooden containers. Incoming materials may arrive in primary and secondary packaging, made from materials such as metal, glass, plastic, ceramic, textile, paper or cardboard, and often on wooden pallets (tertiary packaging). Food safety risk assessment approaches, such as HACCP, should be used to assess each incoming material to identify the potential food safety hazards associated with it and its packaging and the appropriate action necessary to minimise their effects. Preventive measures should start at the source of supply and all raw material specifications should include considerations concerning foreign body control and limitation (see Chapter 3).

Food & Drink – Good Manufacturing Practice: A Guide to its Responsible Management, Seventh Edition.
The Institute of Food Science & Technology Trust Fund.
© 2018 John Wiley & Sons Ltd. Published 2018 by John Wiley & Sons Ltd.

9.3 Internal sources of foreign bodies include the following:

(i) the building and installations (see Chapter 19);
(ii) the plant and equipment (see Chapter 19);
(iii) surface coatings and finishes (see Chapter 19);
(iv) extraneous materials (packaging, cleaning materials and equipment, maintenance, engineering and production tools, spare parts, etc.) (see Chapters 19, 21 and 23);
(v) personnel (see Chapter 17);
(vi) water supply (see Chapter 20);
(vii) pest infestation (see Chapter 22); and
(viii) recovered or reworked product (see Chapter 29).

Prevention 9.4 Preventive concepts should be considered in:

(i) the design of the plant, equipment and buildings and their maintenance;
(ii) the management of non-conforming materials, recovered or reworked product;
(iii) personnel training and management;
(iv) housekeeping and general hygiene; and
(v) processing and packaging.

9.5 The examination and analysis of quality control data and consumer complaints records should be used to monitor the effectiveness of preventive action (see Chapter 27).

9.6 All plant, equipment and buildings should be inspected regularly to ensure that nothing has deteriorated, become dirty or become detached, or is likely to do so, and thereby create a risk of physical contamination of a product (see Chapter 19). Mobile equipment, for example forklifts and pallet trucks, should be used in designated areas, for example external and internal, and be specific to high- or low-care areas in order to minimise the potential for cross-contamination between the different areas established. The Food Standards Agency (FSA) publication *E. coli O157 – Control of cross-contamination: Guidance for food business operators and enforcement authorities revised guidance* (December 2014) focuses on the controls that need to be implemented to prevent cross-contamination.[1] In this context cross-contamination is the transfer of harmful bacteria either directly by food sources coming into contact or indirectly via contact with, for example, a work surface, utensil (e.g. knife) or hands. However, many preventive measures, such as training, sanitation procedures etc., implemented as part of a GMP programme are effective to address not only biological hazards, but also chemical and physical hazards.

9.7 Personnel should be instructed and encouraged to report immediately any incident of contamination/cross-contamination or

[1] https://www.food.gov.uk/business-industry/guidancenotes/hygguid/ecoliguide.

potential contamination of people, materials, packaging, equipment or the product whatever the potential hazard may be. Contamination with a physical item can also lead to biological contamination as the physical item acts as a vehicle to transfer bacteria, e.g. dust, human hair, cleaning cloth, glove and so forth.

9.8   A risk assessment must be undertaken to identify in which areas of the manufacturing site and at which stages of the manufacturing process personnel could present a contamination risk to the product. In areas where personnel are deemed of risk to the product they must be issued with suitable protective clothing which is designed to prevent physical contamination. Overalls should be knee length, have internal pockets only and non-detachable fastenings. Loose items, unless required to carry out necessary work, should be banned in production areas. These include keys, coins, mobile phones and so on. In the event that loose items, for example keys, are required, a 'sign out and sign back in' control protocol should be in place that also identifies the corrective action required in the event an item is lost. All wristwatches, jewellery and bracelets should be prohibited. Plain wedding rings, or wedding wristbands, and secure 'sleeper' continuous loop earrings are possible exceptions if control procedures have been developed to manage these items effectively within the production area. These procedures should also cover rings and studs on exposed parts of the skin such as the face, ears or nose. Fingernails should be clean, short and unvarnished and false nails should not be worn. Adequate head, body and facial hair coverings must be provided and properly used, and should not be kept in place with the aid of hairpins or other fastenings, which could drop off. All head hair should be fully contained in the head covering; sleeves should come to the wrist to prevent the loss of arm hair and potential contamination of the product. Any hoods on personal items of clothing should be under the protective clothing at all times. Clothing with high collars, including those that have zips with metal 'fobs', which could dislodge and fall into the product, should be underneath the protective clothing at all times. Cuts and grazes on exposed skin must be covered by a waterproof dressing, which should be metal detectable, brightly coloured and easily seen against the background of the product. These should be subject to an issue procedure so they are effectively monitored. If disposable gloves are used, they should be checked regularly for any signs of damage or loose pieces that could fall off and contaminate the product. Gloves should be of a distinctive colour, be made of non-allergenic material and must not shed fibres or particles. Gloves should also be subject to an issue control procedure to ensure they are adequately controlled to prevent physical loss and potential product contamination. Smoking, eating and drinking and the use of chewing gum should be restricted to designated areas, away from food production and storage areas, with adequate waste disposal and hand-washing facilities being provided. Employee facilities such as lockers, changing areas, toilets, smoking areas and rest and food areas should be designed and constructed so that they can be maintained in a clean and hygienic manner. The use of clear lockers in food production entry areas aids inspection during hygiene audits. The training programme should explain the necessity for the restrictions and disciplines required in production and storage areas (see Chapter 17).

9.9   Good housekeeping (see Chapters 10, 19 and 21) requires clear instructions concerning the use and disposition of general materials to appear in master manufacturing instructions, plant operating instructions, work instructions, maintenance and service instructions and cleaning manuals. Good housekeeping includes the general tidiness and cleanliness of production and storage areas and also covers infestation control. Buildings must be protected against penetration by animals, birds, rodents and insects by adequate maintenance and proofing. Secondary defences such as poisoned baits, flying insect electrocutors, sticky boards and sprays should be used in appropriate areas to deal with animals and insects that penetrate the building, but should not be seen as the primary means of control. It is generally accepted practice to use non-toxic rodent baits internally within a food manufacturing unit unless a pest infestation has been identified. Due care should be taken to ensure that the infestation controls in themselves do not prove a means of food contamination so they must be sited in suitable locations and securely fastened in place. Wood, glass and paint should, where practicable, be eliminated from open food areas and equipment.

Metal/Foreign Body   9.10   Manufacturing processes should be designed to include process
Detection and   steps and/or procedures that will minimise the risk of foreign body
Removal   contamination of the product. A formal risk assessment should be undertaken to determine whether foreign body detection equipment should be used within the manufacturing process to detect and/or remove foreign body contamination (see Chapter 3). This risk assessment should be undertaken by competent staff and be formally recorded. The frequency of subsequent review of the risk assessment should also be defined and complied with.

Containers may be kept inverted, where practicable, and should be cleaned by jets of filtered air or potable water before being filled. Typical detection and removal activities in a manufacturing process include visual inspection and sorting, air and liquid flotation, spinning, sieving, sifting, washing, filtration, magnets and magnetic grids and plates, metal detection, optical grader/ sorter, and X-ray detection and other forms of scanning.

Effective metal detectors should be employed on the production lines and plant at suitable points in the process. More elaborate methods, such as X-ray examination or scanning equipment, exist and may be useful. In addition to the use of X-ray examination and scanning equipment in factories, cargoes in containers or road vehicles may be X-rayed at some ports during import/

export. Where metal detectors are used within the manufacturing process, they should always include automatic rejection systems and closed containers to hold reject materials. The closed containers should be locked at all times unless being opened by a designated person during monitoring checks. Detection of foreign matter during metal detection may lead to material that can in some instances be reprocessed. This decision on whether to reject or reprocess must be undertaken by suitably competent people who understand the full implications of allowing the product to re-enter the food supply chain. Those responsible for the decision should provide adequate instructions to staff and ensure that suitable monitoring activities are in place.

9.11   Methods of foreign body control should be defined in formal documented procedures. These should include instructions for undertaking the foreign body control procedure, or how to monitor the process step which is designed to eliminate or reduce foreign bodies to a safe level, and the actions to be taken in the event that monitoring identifies product, procedural and/or equipment failure. These foreign body control procedures should address the action to be taken to identify the product that may be affected, its location, the protocol for recall back to the production unit and the procedure for re-inspection. All materials and/or products that have passed through the inspection method or detection equipment since the procedure was last known to be fully operational, that is, working correctly, should be re-inspected. All foreign body detection and removal measures should be validated and re-validated as required (see Chapter 3).

9.12   As previously identified, foreign body controls can include specific process steps such as sieving and filtering. An inventory of filters and sieves should be in place that details the unique equipment identification number, location, type, size in millimetres or microns, and the frequency of inspection. Magnets should also be included on the list if applicable. Sieves and filters should be risk assessed to ensure that they themselves do not present a foreign body risk. This assessment should include reviewing if the sieves are made of metal detectable material or are of a contrasting colour to the food. The review should also ensure the equipment is controlled by documented inspection procedures that identify the action to be taken if the sieves or filters are found to be broken or damaged, the maintenance procedures that have been established for the equipment and the level of training required for operators and those inspecting the equipment. If foreign bodies are found during sieve or filter inspection, then the nature and number of the foreign bodies and the corrective action that has been taken to rectify the problem should be recorded.

9.13   Consideration should also be given to:

• the type of food being analysed, especially particle size and the packaging type (if detection is undertaken following packing). A risk assessment should be undertaken if metal detection is deemed appropriate as to the type of detector and its sensitivity to ferrous, non-ferrous and stainless-steel metal and the degree of sensitivity required. Product packed into metal, metal-laminated or foil packaging should be considered in terms of the alternative metal controls that may be required within the manufacturing process;

- the design of detector required. In-line pipe metal detectors and conveyor-type systems (suitable for large items), which have a belt stop system in place if metal is detected, should be fitted with an audible or visible alarm. Personnel working in the area need to be trained to understand the reasons why such an alarm should go off and the actions that need to be taken. For smaller items, a conveyor-type metal detection system should be fitted with an automatic reject system that has been validated and is routinely monitored to ensure effectiveness. Rejected product should be transferred automatically into a secure, locked box, and there should be an alarm system in place should metal be detected. Personnel working in the area should be aware of the significance of items being rejected and be able to take appropriate corrective action;
- the appropriate mesh/sieve size of filters or sieves;
- the means of validation and the limit of detection of the equipment and whether this represents an acceptable food safety limit. This information will normally be held in the HACCP plan (see Chapter 3);
- the location, that is, where the detector is to be positioned on the process line and whether this location could be bypassed by product in the event of a specific production activity;
- the frequency of monitoring or inspection of the equipment/ method and the verification activities employed to ensure such measures are effective;
- the mechanism for rejection following failure (reject arm, locked box system, automatic line stop). The level of security at such devices must be determined as well as the authority for overriding fail-safe mechanisms, if fitted. Changes to settings or the turning off of reject mechanisms should only be authorised by designated personnel and such decisions must be recorded. Rejection procedures should be formalised, and only authorised personnel should be able to access the locked box or product that has been rejected by the foreign body detector;
- the method of detector calibration (either manual or automatic) as well as servicing and maintenance requirements;
- developing documented procedures that define equipment start-up and operating instructions, and the routine monitoring, testing and calibration of detector equipment, including metal detectors;
- the reporting of incidents and the implementation of effective, timely corrective action. The incident and the corrective action taken should be recorded in the event of a failure of the foreign body detector. This will include stopping the

process, and the subsequent isolation, quarantine and re-inspection of all items produced since the last acceptable test result.

All materials and/or products that have passed through the inspection method or detection equipment since the procedure was last known to be fully operational, that is, working correctly, should be re-inspected. Material detected or removed by the equipment in the case of filters or sieves should be inspected. It should be retained as evidence or, if more appropriate, digital photographs should be taken, so that a full examination can be undertaken and appropriate preventive and/or corrective action adopted.

9.14   Where recycled glass containers are used, provision should be made for the inclusion of automated vision inspection systems to inspect the containers for damage and contamination, including cleaning residues.

9.15   The delivery or storage of materials may involve intermediate packaging to prevent damage. This will subsequently have to be removed. This should be designed to minimise the risk of contaminating the product during its removal. For example, pastry materials are sometimes filled hot into boxes lined with a loose plastic film bag. Creases in the bag become surrounded by pastry that traps the film firmly when it cools. This type of packaging should be avoided, where complete removal cannot be assured. In other cases, if used, internal liners or the film should be highly and contrastingly coloured. Packaging should be clean prior to removal of contents. Materials packed in sealed (seamed) metal containers have the obvious hazard of metal swarf being created when they are opened. Cartons should be staple free; paper sacks should be easy-open, string free and not cut with blunt knives. Plant operatives should be trained to open packaging carefully to avoid product contamination, for example by the misuse of case-opening knives. The use of a deboxing/debagging area with the transfer of ingredients into internal packaging/containers before being transferred to the processing area is recommended to minimise the potential for foreign body contamination. Knife control procedures should be implemented (see 19.43). Other sources of metal to be considered are cutting blades on equipment, chain mail gloves, aprons and arm protection, needles, nuts and bolts, and wires. Control procedures should be in place to monitor metal items in the production area that can become loose and/or damaged, and appropriate corrective action procedures should be in place in the event of such loss or damage.

9.16   All final packaging used by the manufacturer for products should be examined to ensure compliance with the specification against which it is purchased. In addition to this examination, the detailed appraisal of a manufacturing-scale sample, such as a pallet load on a production line, is strongly recommended. This allows the performance of a high-risk packaging material such as glass to be better assessed before acceptance of the bulk delivery for use. In order to implement this protocol, effective lot traceability of packaging material is required. Packaging materials should be brought to their points of usage in minimal quantities (see Chapter 14).

9.17   Brittle material control procedures to control glass, hard plastic or ceramics should be developed (see Chapter 19, 19.36–19.42 check).

9.18   When product containers are cleaned before use, filtered air or potable water should be used. The effectiveness of these cleaning procedures should be routinely monitored and verified, and appropriate action taken based on the results. Where equipment is used to transport, store or hold the product, then the cleaning of these items should be monitored to ensure suitable efficacy. A reject system should be in place to prevent the use of dirty containers that come into direct contact with the product and therefore could present a foreign body risk.

9.19   Where contamination occurs intermittently or infrequently, either systematically or randomly, no practical sampling scheme is likely to detect the fault. Analysis of data produced as a result of monitoring, verification and consumer complaints will indicate any pattern of foreign body contamination and any changes can be subjected to trend analysis that may show the significance or otherwise of the changes (see Chapter 27).

9.20   A risk assessment should be undertaken of offices that open directly into production/storage areas or workstations within production areas. Office equipment, chairs, tables and desks should not be made of wood and should be constructed so that they are easy to clean. Pencils should not be used. Personal items should be kept to a minimum, and no eating or drinking should take place unless water is consumed following an appropriate risk assessment having been made. Stationery items such as paper clips, staples, pens and so on all present a potential foreign body hazard and should be adequately controlled. Quality control workstations should also be risk assessed for the potential for equipment to become lost or damaged and then present a foreign body risk to the product.

9.21   Regrettably, contamination of products by foreign bodies may on occasion be caused deliberately:

(a) during production by an unstable, malicious or disgruntled person;
(b) somewhere in the distribution/retailing chain, by an individual seeking to harm or blackmail a company; or
(c) after purchase, by an individual seeking financial gain or publicity.

While it is difficult to establish complete safeguards against case (a), it is less likely where good industrial relations are fostered. In addition, management should carefully weigh the dangers of allowing particular persons under notice of dismissal or redundancy to work out the period of their notice (see Chapters 5 to 7). Case (b) hazards should be minimised wherever practicable by the use of tamper-evident packaging and tamper-evident seals. As regards (c), careful study of the relevant facts and laboratory examination of the foreign body should be carried out, the results of which may sometimes demonstrate the probability (or even certainty) that it had been introduced subsequent to the pack having been opened.

10   MANUFACTURING ACTIVITIES

Principle

The operations and processes used in manufacture should, with the premises, equipment, materials, personnel and services provided, be capable of consistently producing finished products that conform to their specifications and are suitably protected against contamination or deterioration. Defined manufacturing procedures that address manufacturing operations and associated activities and necessary precautions are required to ensure that all individuals concerned understand what has to be done, how it is to be done, who is responsible and how to avoid mistakes that could affect food safety or quality. For each product, this is provided in the master manufacturing instructions (see Chapter 13).

General

10.1   This chapter deals with those aspects of manufacturing operations and activities that are of general application to any food or drink manufacture. There are, of course, particular additional points and problems concerned with manufacture involving specific types of processing, materials or products. These are the subjects of the chapters in Part II.

Process Evaluation

10.2   Before the introduction of master manufacturing instructions for a product, trials should be carried out to establish whether the formulation and new product brief, and the proposed methods and procedures specified therein, are suitable for factory production and are capable of consistently yielding product that complies with the finished product specification. Validation of food safety process parameters and the effectiveness of prerequisite programmes (PRPs) should also be undertaken at this point and as necessary further amendments and trials should be completed until these conditions are fully satisfied (see Chapters 3 and 12).

10.3   Similar evaluation, and revalidation, should be carried out in connection with any significant proposed change of raw material, equipment or production and/or inspection methods (see Chapters 3 and 12).

10.4   Similar evaluation and review should be carried out periodically to check that the master manufacturing instructions are being followed (see 10.40), that they still represent an effective and acceptable way of achieving the specified product and that they are still capable of consistently doing so (see Chapter 11).

Production

10.5   Adequate resources should be provided in the way of premises, equipment, materials, suitably trained personnel, services, information and documentation, in each case of appropriate quantity and quality (see Chapters 2, 13, 14, 17, 19–24, 26, 28–30, 32–35 and 37–39) to enable the requisite quantity and quality of finished products to be produced.

Food & Drink – Good Manufacturing Practice: A Guide to its Responsible Management, Seventh Edition.
The Institute of Food Science & Technology Trust Fund.
© 2018 John Wiley & Sons Ltd. Published 2018 by John Wiley & Sons Ltd.

10.6 Production should be carried out in full compliance with the master manufacturing instructions from which no departure should be permitted except by written instructions from the production manager and the quality control manager, indicating the nature and duration of the departure, and agreed and signed by them. This is often termed a 'concession'.

10.7 Incentive bonus schemes can create potential hazards and, viewed from the standpoint of food safety and quality, are best avoided. If, however, the provision of an incentive bonus scheme is company policy, it should be so designed as to discourage operators from taking unauthorised shortcuts, for example by building into the formula for bonus calculation a 'quality factor' and/or penalty for observed deviation. In general, prevention of unauthorised short cuts is primarily a task for management through supervision. Where operators have ideas for process improvement, they should be encouraged to raise them (e.g. through suggestions schemes) so that they may be properly evaluated.

Operator Procedures 10.8 Operating instructions for production operators should be written in clear, unambiguous, instructional form, and should form a key part of operator training. Due regard should be given to the reading or language difficulties of some operators.

10.9 Particular attention should be paid to problems that may arise in the event of stoppages, breakdowns or emergencies, and written instructions should be provided for the corrective action to be taken in each such case.

Raw Materials 10.10 Each raw material should comply with its written specification.

10.11 Each delivery or batch should be given a reference code to identify it in storage and processing, and the documentation should be such that, if necessary, any batch (see definition) of finished product can be traced back to the deliveries of the respective raw materials used in its manufacture and correlated with the corresponding inspection, testing, laboratory and/or supplier records. Deliveries should be stored and marked in such a way that their identities do not become lost. Procedures should comply with **EC Regulation No. 178/2002 of the European Parliament and the Council of 28 January 2002** laying down the general principles and requirements of food law, establishing the European Food Standards Authority and laying down procedures in matters of food safety (OJ L 031, 01/02/2002 P. 001 – 0024) and **Article 11, Traceability**, which came into mandatory force on 1 January 2005.[1] Guidance on the general principles and basic requirements for traceability system design and implementation is contained in ISO 22005:2007 (see Chapter 14) and **Commission**

[1] http://eur-lex.europa.eu/legal-content/EN/TXT/PDF/?uri=CELEX:32002R0178&qid=1497376107839&from=en.

Regulation (EU) No. 931/2011 on the traceability requirements set by **Regulation (EC) No. 178/2002 of the European Parliament and of the Council** with regard to food of animal origin.

10.12   Deliveries of raw materials should be quarantined until inspected, sampled and tested in accordance with the quality management system (QMS) and the requirements outlined in Chapter 35, and released for use only on authority of the quality manager, taking account of any certificate of analysis or conformity accompanying a delivery. Particular care should be taken where a delivery of containers appears from markings to include more than one batch of the supplier's production, or where the delivery is of containers repacked by a merchant or broker from a bulk supply. Where appropriate, immediate checks should be carried out for off-flavours, off-odours or taints and, particularly in the case of additives, testing should include test of identity, that is, establishing that the substance is what it is purported to be. (NB: In a multi-container delivery, it is impracticable to check the identity of the contents of every container on arrival, but operators should be trained and encouraged to report immediately anything unusual about the contents when a fresh container is brought into use.)

10.13   Temporarily quarantined material should be located and/or marked in such a way as to avoid risk of its being used accidentally. Material found to require pretreatment before being acceptable for use should be suitably marked and remain quarantined until pretreatment. Material found totally unfit for use should be suitably marked and physically segregated pending appropriate disposal.

10.14   In the case of a bulk delivery by tanker, preliminary quality assessment should be made before discharge into storage is permitted.

10.15   All raw materials should be stored under hygienic conditions and in specific conditions (e.g. of temperature, relative humidity) appropriate to their respective requirements, as indicated in their specifications, and **with due regard to the requirements of legislation on substances hazardous to health, for example, in the UK, the Control of Substances Hazardous to Health (COSHH) legislation**.

10.16   Stocks of raw materials in store should be inspected regularly and sampled/tested where appropriate to ensure that they remain in acceptable condition and compliant with the appropriate raw material specifications.

10.17   When issuing raw materials from store for production use, correct stock rotation should normally be observed, unless otherwise authorised or specified by the quality control manager. This process is very often termed 'FIFO', that is, first-in-first-out stock rotation.

10.18 A formal procedure and associated documentation should be established and followed for the issue of raw materials from store. The personnel responsible for issuing raw materials should be formally defined.

10.19 When a raw material has been issued but not used as planned (e.g. because of a plant (equipment) or production stoppage), the quality control manager or designate should advise as to its disposition. This is critical where items require specific storage conditions such as temperature control in order to maintain food safety or quality.

10.20 Depending on the product being manufactured, the ingredients involved and the nature of the process and equipment, dispensing of the required quantities of ingredients could take various forms, including manual dispensing by weight or volume, automatic dispensing of batch quantities by weight or volume, or continuous metering by volume; the form(s) actually taken will be stated within the master manufacturing instructions. In each case, the weighing and/or measuring equipment should have the capacity, accuracy and precision appropriate to the purpose, and the accuracy should be regularly checked (see Chapter 34).

10.21 Where batch quantities of a raw material have to be dispensed manually into containers in advance, this should be done in a segregated area. Where manual pre-dispensing of relatively small and accurate quantities (e.g. additives) is required, this should be done by, or under direct supervision of, competent staff.

10.22 Records should be kept to enable the quantities of materials issued to be checked against the quantity or number of batches of product manufactured.

10.23 Where an operator controls the addition of one or more raw material to a batch, the addition of each ingredient should be recorded at the time of inclusion on a batch manufacturing record to minimise risk of accidental omission or double addition. Ingredients with key health and safety implications may need to be formally signed off before addition to the batch, for example allergens or preservatives.

Packaging Materials 10.24 Each item of packaging material should comply with its specification (including any legal requirements). The specification should be such as to ensure that:

(a) the product is adequately protected during its expected life under normally expected conditions (with a safety margin for adverse storage);
(b) in the instance of packaging coming into immediate contact with the product, there is no significant adverse interaction between product and packaging material. In the European Union (EU), this includes compliance with the **EU Framework**

Regulation (EC) No. 1935/2004,[2] which defines general requirements for all food contact materials, and a number of specific directives that cover single groups of materials and articles listed in the framework regulation, as well as directives on individual substances or groups of substances used in the manufacture of materials and articles intended for food contact;

(c) where the packaged product undergoes subsequent treatment, whether by the manufacturer, caterer or consumer, the packaging will adequately stand up to the processing conditions and no adverse packaging/product interaction occurs;

(d) the packaging is capable of providing the necessary characteristics and integrity where the preservation of the product depends on the pack; and

(e) the finished pack will carry the statutory and other specified information in the required form and location.

10.25 Where packaging material (e.g. labels, printed packages, lithographed cans) carries information required by law, the quality control manager should ensure that the specification is updated as required to comply with new legal provisions. In some organisations this responsibility may be within the product development department. The quality manager should ensure that stocks of packaging materials that no longer comply are quarantined for modifications (if possible and desired) or destruction. Similar provisions and precautions should apply in the case of contract packing for a customer where the latter requires changes to other label information (see Chapter 33). The requirements under good manufacturing practice (GMP) for smart packaging, including active and intelligent packaging, are included in Chapter 25.

10.26 Each delivery or batch of packaging should be given a reference code to identify it in storage and processing, and the documentation should be such that, if necessary, any batch of finished product can be traced to deliveries of the respective packaging materials used in its manufacture and correlated with the corresponding laboratory records. Deliveries should be stored and marked in such a way that their identities do not become lost (see Chapters 14 to 16).

10.27 Deliveries of packaging material should be quarantined until inspected, sampled and tested in accordance with the QMS and Chapters 28 and 30, and released for use only on authority of the quality manager. Deliveries of packaging should be adequately inspected to confirm that they comply with the relevant specification and that the external packaging is intact and unopened. Damaged, previously opened or dirty cartons of packaging must not be accepted due to the risk of contamination. Operators should be trained and encouraged to report immediately anything

[2] http://eur-lex.europa.eu/legal-content/EN/TXT/PDF/?uri=CELEX:32004R1935&from=en.

unusual about the appearance, odour or behaviour of packaging materials issued, especially colour differences, changes in print quality or problems with sealing.

10.28   Temporarily quarantined packaging material should be located and/or marked in such a way as to avoid risk of its being accidentally used before release. The personnel responsible for the decision on its disposition should be formally defined. Material found totally unfit for use in packaging operations should be suitably marked and physically segregated pending appropriate disposal.

10.29   All packaging materials should be stored in hygienic conditions and as indicated in their respective specifications.

10.30   Stocks of packaging materials in store should be inspected regularly to ensure that they remain in acceptable condition, and appropriate infestation/pest control procedures must be in place for storage areas (see Chapter 22).

10.31   In issuing packaging material from store for production use, stock rotation should normally be observed, unless otherwise authorised or specified by the quality control manager. This process is very often termed FIFO stock rotation.

10.32   A formal procedure and associated documentation should be established and followed for the issue from, and the return of, part-used batches of packaging to store. The procedure should address the need to reseal part-used boxes of packaging to prevent foreign body contamination. The personnel responsible for issuing and reconciling packaging stocks should be formally defined.

10.33   Where a company manufactures more than one product, or more than one version of a single product, the greatest care should be taken to check that the correct packaging is issued for the product to be manufactured, and that no incorrect packaging materials left over from a previous production run of a different product or a different version are left in the production area, where they might accidentally be used. This is especially important when products are placed on promotion and the packaging specification is amended, and then after a given duration the product is packed in the original packaging. Particular attention should be paid to controlling adhesive, flash and price labels, especially where the design of the labels is very similar.

10.34   Under no circumstances should primary food packaging be used for anything other than its intended purpose.

10.35   Where packaging is reference coded, priced or date marked for use, care should be taken to ensure that only material carrying the correct production date is used. Surplus material left from earlier production runs and no longer bearing a valid reference or date should not be left in the production area. Where the reference and/or date is applied during the manufacturing operation, care should be taken to check and ensure that the marking machine is set for the correct reference and date. Where packaging or labels are generated in an off-line print room, procedures need to be in place to sign off/verify that the print details are correct when released from the print room and when accepted by the production line. A protocol should be in place to ensure that any unused packaging or labels are suitably controlled so that they cannot be inadvertently used in production. Label approval, monitoring and verification activities must be formally defined and appropriate training given to those personnel assigned these tasks so they understand the meticulous inspection required and the consequences if an incorrect label is not identified and used within production. Crisis management, product recall and product withdrawal procedures must be in place that address the procedures to be followed in the event that incorrect product labelling has occurred, especially if the product is known to, or may potentially, contain an allergen that is then not identified on product labelling/packaging (see Chapter 27).

Processing and Packaging

10.36 Where a company manufactures more than one product or more than one version of a product, and there is more than one production line, production layout should be such that confusion and possible cross-contamination are avoided (see 9.6).

10.37 Whether in single-line or multiple-line production, particular care should be taken in terms of production layout and practices to avoid cross-contamination of one product by another (see 9.6).

10.38 On a production line, in order to avoid confusion, the name and appropriate reference to the product being processed or packaged should be clearly displayed or otherwise communicated.

10.39 Before production begins, checks should be carried out to ensure that the production area is clean and free from any products, product residues, waste material, raw materials, packaging materials or documents not relevant to the production to be undertaken, and that the correct materials and documents have been issued and the correct machine settings have been made. All plant and equipment should be checked as clean and ready for use. A formal form or checklist may be designed for this purpose.

10.40 Processing should be strictly in accordance with the master manufacturing instructions subject to any variations approved (as in 10.6), and by detailed procedures set out for operators in the plant operating instructions. Master manufacturing instructions, product or process specifications should define the following:

(a) pre-startup, close-down and product changeover checks to ensure the line is cleared for production and closed down correctly at the end of the shift;

(b) the product recipe in terms of ingredients and proportions;

(c) mixing instructions, including machinery to be used, duration of mixing and processing requirements;

(d) equipment settings and processing parameters (time, pressure and temperature);

(e) the control of in-process monitoring devices and divert systems if minimum processing parameters are not reached;

(f) filling requirements in terms of volume or weight;

(g) criteria for in-process foreign body inspection or detection;

(h) inspections to ensure processing conditions are consistent throughout the batch being processed;

(i) packaging and labelling requirements, especially at start-up and product changeover;

(j) sampling criteria in terms of parameters to be checked and the frequency of sampling; and

(k) additional food safety and quality criteria.

10.41 Process conditions should be monitored and process control carried out by suitable means, including, as appropriate, sensory, instrumental and laboratory testing, and on-line checking of correct packaging and date marking. Where checkweighers, filler systems, continuous recorders or recorder/controllers are in use, the data produced should be checked by quality control and charts or printouts retained in either electronic or paper form as process records. This is particularly important at process steps that are deemed by the manufacturer as a critical control point (CCP) for a food safety hazard (see Chapter 3).

10.42 There should be regular and recorded checks by appropriate personnel on the accuracy of all instruments used for monitoring processes (e.g. temperature probes/gauges, pressure gauges, flow meters, checkweighers, colour measurement devices, metal detectors and X-ray machines). The frequency of checks should be established to ensure that instruments are always correctly calibrated, with an accuracy related to national standards (see Chapter 34).

10.43 Effective cleaning of production premises and equipment must be carried out (see Chapter 21).

10.44 All persons working in or visiting the production area must comply with the requirements of personal hygiene and adequate facilities must be provided, as summarised in Chapter 17.

10.45 General 'good housekeeping' should be practiced, including prompt removal of waste material, precautions to minimise spillage or breakage, prompt removal and clean-up of any spillage or broken packaging, and the removal of any articles that might enter the product as foreign matter, otherwise known as a 'clean-as-you-go' policy (see Chapter 21).

10.46 Where appropriate, foreign matter detectors should be used (see Chapter 9).

Intermediate Products

10.47 After its preparation, an intermediate product should be held until checked and approved by quality control for compliance with its specification. If required to be stored before further processing, it should be stored as designated in the specification, and suitably reference marked and documented so that it can be traceable and correlated with the lots of the raw materials from which it was made and the batch(es) of finished product in which it is subsequently incorporated (see Chapter 14).

10.48 A batch of intermediate product found to be defective should be quarantined pending re-working or recovery of material or outright rejection as the case may be. The personnel responsible for the decision on its disposition should be formally defined (see Chapter 29).

Finished Product

10.49 Packed finished products should be quarantined until checked and approved by quality control for compliance with the appropriate finished product specification.

10.50 An approved batch of finished product should be suitably marked to identify it, and stored under the appropriate conditions (e.g. of temperature or relative humidity) stated in the finished product specification. Under the traceability requirement of **EU Regulation No. 178/2002**, an identification mark must be provided linked to a traceability document to any retailer to whom part of that batch is sold.

10.51 Where a batch of finished product fails to meet the specification, the reasons for failure should be thoroughly investigated. Defective finished product should remain quarantined pending re-working or recovery of materials or disposal as the case may be. The personnel responsible for the decision on its disposition should be formally defined (see Chapter 28).

Storage

10.52 Desirable features of storage premises are outlined in Chapters 19 and 26. Aspects of storage of goods in the distribution system are dealt with in Chapter 32.

Transport

10.53 Materials or products should be transported within the factory premises in such a way that their identities are not lost; that there is no mixing of materials or products approved for use or despatch with those that are quarantined; that by-products, particularly those not intended for human food use, do not lead to contamination of other materials; that spillage is prevented; no breakage or other physical damage is caused to the goods being transported; and that goods being transported are not left in adverse conditions or otherwise allowed to deteriorate. Aspects of transport of goods in the distribution system are dealt with in Chapter 32.

11 MANAGEMENT REVIEW, INTERNAL AUDIT AND VERIFICATION

Principle

Management review is the process of reviewing the quality management system (QMS)/food safety management system (FSMS) and food integrity management system (FIMS) and is undertaken by senior management at regular intervals. Management review is undertaken to determine the degree of management control and its effectiveness, and verify that the food safety and quality policy and food safety and quality objectives, QMS, FSMS and FIMS, are suitable and being complied with. It is also an opportunity to identify any areas for improvement. An internal audit programme should support the management review process. It provides an input to management review by measuring the level of conformance with the QMS/FSMS/FIMS and the effectiveness of the QMS/FSMS/FIMS in achieving food safety, legality and quality objectives.

Management Review

11.1 Management review is formally addressed by manufacturing organisations at regular structured management meetings. The effectiveness of the management review meetings requires certain prerequisites to be in place, including determining:

- who will be on the senior strategic management review team and any operational management review teams that might operate at the department level;
- who will chair the meetings;
- how often the meetings will take place;
- the agenda that will be followed;
- who will take and maintain the minutes of the meeting; and
- the identification of appropriate preventive and corrective actions that arise from discussions at the meeting.

The meeting minutes form part of the formal records of the organisation (see Chapter 13) and need to be retained to provide objective evidence that the meeting(s) took place The minutes also need to identify the actions that were agreed as a result of the meeting and any follow-up action that was agreed at the meeting to ensure that the agreed actions were implemented and were effective. Clear actions need to be agreed with responsibilities for action and timescales for completion. It is important that all actions are SMART (specific, measurable, achievable, relevant and time based) otherwise preventive or corrective action will be limited in its degree of effectiveness. The channels of communication of the meeting decisions and the actions that have been agreed must be formalised. The frequency of meetings can be monthly, bimonthly or quarterly. If the meetings are any less frequent, then they become a historic review rather than a real-time management process that drives continuous improvement.

Food & Drink – Good Manufacturing Practice: A Guide to its Responsible Management, Seventh Edition.
The Institute of Food Science & Technology Trust Fund.
© 2018 John Wiley & Sons Ltd. Published 2018 by John Wiley & Sons Ltd.

For the same reason, minutes need to be circulated to those concerned without delay as a guide to agreed decisions and in order to facilitate prompt action.

Inputs to the Review Process   11.2   Inputs to the management review process can include the following:

- progress in complying with previous management review meeting minutes and action plans;
- areas of non-compliance, including incidents, customer complaints, non-conforming items and status of existing corrective actions as well as the effectiveness of previously completed corrective actions;
- progress in complying with the QMS, FIMS, FSMS and hazard analysis critical control point (HACCP) plan and any non-conformance and subsequent corrective action;
- progress in complying with food safety and quality objectives and trends in food safety and quality costs such as level of rejection, rework and cost of customer complaints;
- progress in complying with internal food integrity plans such as security assessments, vulnerability assessments and threat analysis critical control point (TACCP) plans or their equivalent (see Chapter 6);
- horizon scanning for potential supply issues which could mean that there is a greater risk to food integrity, safety or quality, e.g. poor harvests, weather events such as drought in key supply locations or social issues that will impact on supply;
- preventive actions and opportunities for improvement that have been or need to be addressed;
- results from internal, external and third-party audits;
- changes to the documented system since the last review;
- general customer feedback and analysis of customer and/or consumer complaint data including service levels and subsequent investigations and actions taken;
- changes to the organisational structure or individual job responsibilities;
- analysis of supplier performance and approval of new suppliers and contractors;
- analysis of product and service performance;
- changes that could affect the QMS/FSMS/FIMS system, including new market requirements, legislation, technology, products and processes; and
- training programmes and the current training needs/plans, including effectiveness of previous training.

11.3   If an integrated management review process is undertaken, inputs could also include social accountability and organisational and food safety culture, animal welfare, corporate social responsibility, personnel health and safety, and environmental issues.

Outputs from the Review Process   11.4   Measurable outputs that demonstrate the effectiveness of the management review process include:

- continued compliance with statutory, legislative and market requirements;
- improved allocation of resources;
- continued focusing of the food safety and quality policy and food safety, integrity and quality objectives on the issues affecting the manufacturing organisation;
- improved planning and communication on future changes within the business, including new products, technology and processes, new suppliers, changes to the organisational structure and individual job responsibilities or the documented system;
- improved management of the corrective and preventive action programme;
- monitoring of supplier performance and improved communication with regard to issues of service and raw material/ingredient performance;
- monitoring of the status of food integrity, safety and quality risks;
- improved relationships with customers;
- measurement of food safety, integrity and quality costs on an ongoing basis towards reduction in non-conforming product and service; and
- continued focusing of training programmes and identification of current training needs.

Resourcing the QMS/FSMS/FIMS

11.5 It is the role of the senior management team, using the management review process as a driver, to provide the resources required to fulfil food safety, food integrity and quality objectives and to ensure legal compliance. Management review needs to consider whether existing human, physical and financial resources are adequate to achieve this and, where this is not the case, drive the implementation of appropriate corrective action so that adequate resources are in place.

Legislative and Industry Guidance

11.6 The senior management review process should have a formal process in place to ensure that the manufacturing organisation can access, and then address within its management systems and procedures, scientific and technical developments, including the emergence of new food safety hazards, food quality issues or food integrity concerns and mitigation measures, changes and updates to industry guidance and codes of practice, and changes to legislation both in the country of manufacture and those countries to which the organisation is seeking to export its products.

Types of Audit

11.7 Auditing is the management tool that identifies whether the food safety, integrity and quality manual, and associated procedures and systems, including prerequisite programmes (PRPs), have been developed and implemented appropriately, are effective and are being complied with and that there are no weaknesses in the formal system that could give rise to non-conformance and/or no evidence of actual non-conformance evident during the audit. Audits can be described as:

- first-party or internal, where the organisation is auditing its own systems and procedures;
- second-party, where the organisation is undertaking audits of organisations with whom it has a contractual agreement, e.g. suppliers. The customer, e.g. a food retailer or a food service organisation, will very often develop their own standard against which they are auditing; or
- third-party, where an independent organisation is auditing typically a supplier of a retailer or food service organisation and there is no contractual product supply agreement between them and the organisation that they are auditing. Examples of third-party audit standards include the British Retail Consortium Global Standard Food Safety or the ISO series of system standards.

11.8 Audits can be undertaken to two levels of depth: the first is a *system audit* where the auditee's (the organisation being audited) documented management system is audited against the system requirements of the standard, i.e. how the documented management system should be constructed and what elements it should contain. The second level of audit is a *compliance audit*, which measures if what the manufacturing organisation is doing meets the requirements of their own documented management system and the requirements of the system standard, e.g. the organisation's documented management system may have greater requirements in place than the system standard, e.g. with regard to training, traceability or provenance standards. It is also important to determine the scope of the audit, i.e. the products and processes that will be audited and identified as being compliant, or not, in the audit.

Internal Audits 11.9 Internal audits complement the management review process. Internal, or first-party, audits are where the organisation develops an auditing programme to audit itself in terms of products, procedures and processes. The auditing programme should include all the activities within the scope of the QMS/FSMS/ FIMS. The audit programme should be defined in an audit schedule that defines audit criteria, scope, the proposed auditor and planned frequency of audits. An internal audit procedure should be developed defining the requirements for the internal audit programme, including planning and conducting audits, mechanisms for reporting the results of the audit and how any required actions will be monitored and followed up. Individual responsibilities should be defined within the procedure. The procedure should also outline the working documents that will be used, such as audit reports, audit checklists and corrective action plans. The resource required for the internal auditing programme should be reviewed, including the required number of auditors and the training required in order to develop the auditors' specific auditing skills and technical knowledge. The resource will depend on the number of audits scheduled and their scope and frequency. The frequency of audits should be established by a formal risk assessment. The internal audit procedure should identify the depth of the internal audits, i.e. whether they are system or compliance audits, or both, and the scope of the audit, i.e. not only the area of activities being audited, but also the elements of the management systems (s) (QMS, FSMS and/or FIMS and also environmental, health and safety of personnel or other criteria) for a compliance audit and the system standard such as the British Retail Consortium (BRC) Global Standard for Food Safety, alternative private or retailer standard, ISO 22000 or EN ISO 9001:2000 or other relevant standard in the event of a system audit. Auditors should have technical knowledge of the food products and processes being audited. Auditors cannot objectively audit their own work, so internal audits should be carried out by competent auditors who are independent of the area or activity being audited. This will ensure the objectivity and impartiality of the audit process.

11.10   The results of the audit should be documented and brought to the attention of the management responsible for the area being audited. This should include areas of both conformance and non-conformance. Any preventive or corrective actions and timescales for their implementation should be mutually agreed. The management responsible for the area being audited is responsible for ensuring that any required actions are undertaken in a reasonable time frame in order to eliminate non-conformance and ensure effective corrective action. Verification activities should then be undertaken to ensure that the prescribed preventive or corrective actions have been implemented and have been effective. A process should be put in place to manage all preventive and corrective action required within the organisation. This can be through a preventive and corrective action plan, non-conformance or corrective action log or similar document. The quality control manager should be responsible for monitoring the completion of preventive and corrective action according to agreed timescales and should highlight poor performance for follow up, additional resources and/or further action.

11.11   The implementation of quality assurance processes within an effective QMS/FSMS/FIMS requires not only the implementation of procedural audits but also those that address extrinsic food safety hazards and food integrity threats. This requires an audit plan to be established to monitor the 'fabric' of the manufacturing premises in terms of the building and its physical security, zoning and access to specific areas within the manufacturing premises, use and security of equipment and tools, premises housekeeping and hygiene, and personnel hygiene. The frequency of these audits should be based on risk assessment, including assessing the degree to which the product is enclosed within food-processing equipment or secure production zones. These premises and personnel audits should be undertaken by trained, competent individuals, and where possible should not just be seen as simply as the completion of a checklist and an opportunity to identify non-conformance, i.e. a tick-box activity. Auditors should be encouraged not only to have awareness of the premises and personnel standards required but also to use the internal audit process to improve staff understanding of the good manufacturing practice (GMP) standards required and promote continuous improvement. Senior management should support this activity as a preventive approach to minimising food safety and food integrity risk.

11.12   Trend analysis should be undertaken across a series of internal audits to identify areas that give rise to ongoing, albeit minor, non-conformance, especially where corrective action is shown to have limited value in addressing weaknesses or breakdowns in the QMS, FSMS and/or FIMS. This should form an input into the management review process, as previously described. The trend analysis should form a framework for the development of specific key performance indicators (KPIs) or critical success factors (CSFs) that are robust enough to drive continuous improvement in all aspects of GMP.

11.13   Guidance on undertaking audits and developing auditing programmes can be accessed in **BS EN ISO 19011:2011** *Guidelines for auditing management systems*. Although not relevant to internal auditing itself ISO/TS 22003:2013 *Food safety management systems – Requirements for bodies providing audit and certification of food safety management systems* provides guidance relevant to the undertaking of audit activities that still proves useful in developing an internal auditing programme. The annexes of this standard provide specific guidance on the time requirements for auditing.

11.14   Guidelines for auditing FIMS in the food manufacturing environment are limited to date. However, when considering the development of an auditing programme to verify the four aspects of food integrity (product, process, data and people) it is important to consider the ISO/IEC 27000 standards for information security management systems which are under development at the time of writing this Guide, including:

- ISO/TEC 27003:2017 *Information technology – security techniques – information security management systems – guidance*;
- ISO 10667-1:2011 *Assessment service delivery – Procedures and methods to assess people in work and organizational settings*;
- ISO 31000:2009 *Risk management – Principles and guidelines*.

12   PRODUCT AND PROCESS DEVELOPMENT AND VALIDATION

Food & Drink – Good Manufacturing Practice: A Guide to its Responsible Management, Seventh Edition.
The Institute of Food Science & Technology Trust Fund.
© 2018 John Wiley & Sons Ltd. Published 2018 by John Wiley & Sons Ltd.

Principle

Products should be designed to ensure that the end product meets consumer expectation within the intended and anticipated circumstances of use, and to ensure that product design and performance have been fully evaluated for the required function in respect of microbiological safety, chemical safety, physical safety and sensory quality as elaborated in Chapter 2. To this end, a hazard analysis critical control point (HACCP) study undertaken by a multidisciplinary team is recommended, preferably applied from the earliest stages of product or process development with a view to eliminating or minimising potential hazards wherever possible and incorporating effective control parameters in the product design (see Chapter 3). The HACCP study should encompass any possible deviations from the design objectives. It is important to consider all realistic potential hazards that may occur at each stage of the manufacturing process. In order to design the product to achieve these objectives, it is necessary to identify all the possible causes of the identified deviations. A hazard analysis and operability (HAZOP) study is a recognised hazard analysis method used during process development and validation to identify both food safety hazards and potential operational issues that could lead to food safety or environmental hazards or impact on manufacturing efficiency (see Chapter 3). Similar considerations apply where changes are made to existing products that would affect the integrity, safety or stability of a product. Such changes could include those made to:

(a) *ingredients or supplier;*
(b) *formulation or recipe;*
(c) *operations, e.g. order of addition;*
(d) *machinery;*
(e) *processes or process parameters;*
(f) *packaging;*
(g) *storage;*
(h) *distribution;*
(i) *organisational structure or staff responsibilities; or*
(j) *consumer use.*

The above list is not exhaustive but is based on experience of situations where failure to take such changes into account has resulted in serious human or commercial consequences or both.

Customer Requirements

12.1   Before commencing the design procedure, it is advisable to define the customer and/or consumer expectations of the new product or process. These should be established by market research or by direct agreement with the customer (e.g. the retailer, caterer or distributor). If established by market research, these expectations should be translated clearly into design objectives. It is also important to consider whether the food is targeted at a section of the population that is deemed as a vulnerable group with regard to food safety, for example the elderly, the young, pregnant women or immunocompromised individuals, and the possible implications in terms of product design. Customer requirements may include the utilisation of recognised food safety hazards, for example allergens in the product formulation, a formulation, packaging design or manufacturing process that may lead to higher risk of pathogen survival, or the use of packaging that in itself presents a food safety hazard, for example glass, brittle plastic or metal. Some products or food ingredients may be more vulnerable to integrity issues, such as very expensive spices that can potentially be substituted with inferior material and/or other ingredients that are liable to supply shocks caused by drought, weather impact etc. It should be noted at the start of the product development process if there is the potential for a risk of loss of integrity that cannot then be appropriately mitigated.

12.2   Due account should be taken of any specific standards that are to be met to justify special claims (including those related to packaging), specified requirements such as suitability for a specific nutritional or health purpose, or other particular expectations, for example:

- provenance and geographic location;
- environmental benefits, e.g. recycling,
- use of novel technology;
- claims of being free from specific ingredients, e.g. sugar free;
- specific production method: halal, kosher, organic, vegan, vegetarian;
- animal welfare friendly;
- diabetic or other health-related product; or
- low-carbohydrate, low-fat or suitable for a low-calorie diet.

Those claims that could be attributed to the product should be agreed before beginning the design process.

Design Objectives   12.3   Before commencing the design of a new product, design objectives should be established clearly, having due regard to the following requirements:

(i)   customer/consumer expectations;
(ii)   product safety;
(iii)   product integrity, including the need for authenticity markers;
(iv)   regulatory requirements for the intended market; and
(v)   minimising operating costs while remaining consistent with the fulfilment of (i), (ii) and (iii).

Design objectives should include acceptance criteria and tolerances. Establishing tolerances is crucial to achieving customer expectations at a realistic cost. They may be established by past precedent or by agreement with the customer (e.g. the retailer, caterer or distributor).

Regulatory Requirements	12.4	Where regulatory requirements include statutory maxima or minima for particular parameters, the design targets should provide at least sufficient margin below the maxima or above the minima to allow for ingredient composition variation, within and between batch variation and process variation.
Planning	12.5	Planning for a new process or product should include clear identification of responsibilities for the design activities, agreed actions and timescales. This is best achieved through a multidisciplinary project team, which could comprise technical, production, engineering, marketing, and purchasing and distribution expertise. In the event of changes to existing products, packaging or processing methods, the existing HACCP plan(s) and food safety management system (FSMS), quality management system (QMS) and food integrity management system (FIMS) needs to be reviewed by the HACCP team and/or project team if they are the same group of individuals. Any proposed changes should be validated. Validation should include a range of trials as applicable to the situation (see 12.6). Validation is the collection and evaluation of 'scientific and technical information to determine whether the HACCP plan, when properly implemented, will effectively control the identified food hazards'.[1]

When a new product is proposed, the HACCP team should consider whether the product is addressed by an existing food safety/ HACCP plan, or whether a new HACCP plan needs to be initiated. In the event of a new packaging or process system being introduced, a new HACCP plan needs to be developed. New HACCP plans or a modified existing HACCP plan must be approved by the HACCP team before factory trials can take place.

Quality, Safety and Integrity Planning	12.6	At each critical stage in the design process (e.g. kitchen, pilot scale, pre-production factory trials, first production, packaging trial, distribution (transit) trial, retail trial) it is essential to verify that the product meets its agreed design criteria and agreed tolerances, and that the HACCP study has been successful in effectively managing food safety hazards. This can be achieved in a number of ways, including:

- shelf-life testing;
- quality control testing of microbiological, chemical and sensory parameters and packaging;
- market trials;
- process monitoring; and
- checks for compliance with regulatory requirements.

The trials should be undertaken in accordance with documented procedures and the results of the trials should be documented and

---

[1] Food Standards Agency (2014). *E. coli O157 – Control of cross-contamination: Guidance for food business operators and enforcement authorities.*

retained. If the factory trials and associated testing form part of the validation process for the HACCP plan and general FSMS and FIMS, these should be referenced within the HACCP documentation. The FSA publication *E. coli O157 – Control of cross-contamination: Guidance for food business operators and enforcement authorities* (2014) stresses the importance of not only validating the HACCP plan, but also focusing on validating the control measures in place to ensure that bacterial loading on fresh produce is reduced on receipt, that physical separation of materials is effective and that disinfectants are purchased and used in compliance with validated dilution levels and contact times.

Traceability of materials used in factory trials is critical, and trial ingredients, in-process material and finished product should be suitably labelled, segregated and controlled to ensure that it does not enter the general stock system. If the materials used are, or contain, potential allergens, then allergen-control procedures should be followed (see Chapter 8).

12.7   Nutrition analysis should be undertaken to ensure that at the end of the shelf life, nutrient values comply with the product label, especially where the original design phase involved calculation of nutrition content from standard sources. This is critical where nutritional claims are made for the product, for example 'free from', 'high in' and 'low in' claims.

12.8   A food crime risk assessment needs to be undertaken to determine the potential vulnerabilities associated with the new products and any potential integrity risk, or areas where transparency is weak or there are potential threats that need to be adequately controlled (see Chapters 5 to 7).

12.9   A quality plan needs to be developed to manage key product quality parameters. Guidance on the development of quality plans can be accessed in ISO 10005:2005 *Quality management systems – Guidelines for quality plans*.

12.10   Failure to achieve the design objectives can be caused by three factors: unrealistic initial design objectives, an inadequate HACCP study, risk assessment or quality plan, or failure to implement the requirements properly within the product development process. In all cases the entire new product development procedure should be revisited, although revision of the initial objectives and design criteria may be considered if the end result is demonstrably acceptable to the manufacturer and their customers in all respects despite the failure to meet an original design objective that is no longer deemed necessary. If the objectives are changed at any point during the design and development process, a thorough re-evaluation of the implications of the changes in terms of food safety, integrity and quality must be made.

Labelling   12.11   Checks should be undertaken to ensure that the labelling of the new product is legal and in accordance with the formulation and product specification. These checks should include weight control and volume control information, nutrition information, allergen information, customer storage instructions, cooking and preparation methods and other instructions for use, and product durability information (see Chapter 37).

13 DOCUMENTATION

Principle

Effective documentation is an essential and integral part of good manufacturing practice (GMP) and, in particular, one of the essential features of a properly operated quality management system (QMS), hazard analysis critical control point (HACCP) plan, food safety management system (FSMS) and food integrity management system (FIMS). The purposes of formalised (documented) systems are to define the scope of the management systems, the materials, operations, activities, control measures and products; to record and communicate information needed before, during or after manufacture; to reduce the risk of error arising from oral communication; and to develop records that form a vital part of the audit trail necessary for tracing components used to produce the final product. The system of documentation should be such that as far as is reasonably practicable the history of each batch of product, including utilisation and disposal of raw materials and packaging, intermediates and bulk or finished products, may be ascertained.

General

13.1   Documents fall into three main categories:

(a) manuals, which define company strategy, policy and protocols and how the organisation may seek to address a QMS standard such as ISO 9001:2015 *Quality management systems – Requirements*, or the British Retail Consortium (BRC) Global Standard for Food Safety;

(b) procedures, which define the company prerequisite programmes (PRPs), countermeasures and system procedures such as customer complaint procedures, management of incidents and product recall, traceability or calibration procedures; and

(c) work instructions, standard operating procedures, specifications and master forms. These are the documents that define production activities and identify the quality standards and control checks that need to be undertaken. A master form is the blank template, which once completed becomes a quality record. These records can often form part of the organisation's due diligence documentation.

Failure to undertake the required activities and complete the appropriate records will nullify many of the benefits of GMP, food integrity management or effective manufacturing operations and effective food control. Documents may include the following types.

(a) **Manuals and policies, including:**
- quality manual;
- HACCP manual/plan;
- health and safety manual;

Food & Drink – Good Manufacturing Practice: A Guide to its Responsible Management, Seventh Edition.
The Institute of Food Science & Technology Trust Fund.
© 2018 John Wiley & Sons Ltd. Published 2018 by John Wiley & Sons Ltd.

- Control of Substances Hazardous to Health (COSHH) manual (in the United Kingdom (UK) or equivalent elsewhere);
- quality policy;
- food safety policy;
- food integrity policy;
- allergen policy;
- environmental policy;
- ethical trading policy; and
- health and safety policy

(b) **Procedures and programmes, including:**
- production programmes;
- training programmes;
- internal quality audit programme;
- pest control programme;
- plant maintenance programme;
- procurement procedure;
- supplier approval and performance monitoring programme;
- management of incidents and product withdrawal and recall procedure;
- quality control (including analytical and microbiological) procedures and methods;
- premises housekeeping and hygiene, including cleaning schedules;
- traceability procedure;
- calibration procedure;
- personnel hygiene procedure;
- document control procedure; and
- quality records management procedure, including archiving.

(c) **Work instructions, standard operating procedures, specifications and master forms such as:**
- ingredient specifications;
- packaging materials specifications, including artwork standards;
- copy of order and/or terms of conditions of purchase;
- master manufacturing instructions, job notes and material resource planning instructions, including flowsheets and standard recipes;
- work instructions and task procedures;
- product formulations;
- intermediates specifications;
- bulk products specifications;
- finished products specifications; and
- machinery operating instructions.

(d) **Records and reports (system records)**

These documents in particular should be designed with due regard to intelligibility to the intended user, including the need to recognise possible numeracy, literacy and language problems. These documents include:
- records for monitoring of critical control points (CCPs) and critical quality points (CQPs);

- records of receipt, examination, approval and issue for use of raw materials and packaging materials;
- records of the testing and release of intermediates, bulk products and finished products;
- records of process control tests;
- in-process recording instrument charts;
- weight or volume control charts;
- batch manufacturing records;
- customer complaint and service level reports;
- minutes of management system review meetings;
- quality control summaries and surveys;
- quality assurance, audit reports and records; and
- superseded documents.

(e) **Additional system documentation, including**
- site plans, e.g. water system plans, plant layout, drainage plans;
- copies of applicable codes of practice and relevant legislation; and
- equipment manuals.

13.2 Detailed advice on documentation related to specific production categories and specialised topics are contained in many of the chapters of this Guide. These chapters should be consulted in conjunction with this chapter.

Quality Manual 13.3 A quality manual should outline the structure of the QMS and provide a means to reference associated system procedures and programmes, work practices, policies and forms. The quality manual must be established and implemented, and verification activities such as internal audits (see 11.9) must be undertaken to ensure its appropriateness to the processes undertaken and the degree of effectiveness in addressing food safety and quality criteria. The quality manual should be available to all relevant staff. It should integrate fully with the HACCP documentation (and, as appropriate, threat analysis critical control point (TACCP) and vulnerability assessment plans and documentation; see Chapter 6) and define how the organisational structure has been developed to ensure food safety, food integrity and product quality, and continued compliance with defined product and process criteria.

Formal Policies 13.4 Formal policies should be signed by the member of the senior management team with overall responsibility for the safety, legality, integrity and quality of the products manufactured. A *quality policy* is a statement of the overall mission, intentions and strategic direction of an organisation as related to quality criteria. The quality policy must be documented and approved by senior management. The policy defines the organisation's intentions and commitment to manufacturing safe and legal products that meet customers' quality expectations. The organisation must demonstrate that it has communicated its quality policy to all personnel and that they understand the importance of the policy statement and associated QMS.

A food safety policy addresses the food safety intentions and objectives of the business and can be a standalone document or be integrated with the quality policy into a joint statement. Equally the quality manual may be a standalone manual that interfaces with the HACCP plan or a combined food safety and quality manual.

A food integrity policy may be a standalone document or subsumed into an integrated food safety, quality and integrity policy. A food integrity policy statement addresses the intentions of senior management to ensure that there is accountability with regards to products, processes, data and employees' activities and actions. Quality, safety and integrity objectives should be formalised, that is, documented, time based and measurable, so that achievement can be demonstrated. Policy objectives should be communicated to all staff, whether through the formal policies or other means. Consideration should be given to the reading or language difficulties of staff when these objectives are communicated. Appropriate personnel, at a minimum time frame of annually, should monitor the ongoing progress towards the achievement of food safety, integrity and quality objectives, and the results form an input into the management review process (see Chapter 11).

Delivery/Intake Inspection Records   13.5   Due to the requirements of the supplier and buyer, and the need for inspection records containing certain information along with other recorded data, a separate record is normally required for the inspection of every delivery or part thereof. Copies of this record may go to the vendor, production manager, purchasing manager and quality control manager. This documentation will also play a pivotal role in the case of contract manufacturing (see Chapter 33). Considering the number of times this information is used, it is necessary to make the process convenient and brief, but complete in all the essential details relating to the consignment. This record will also be an important link in effectively managing product identification and traceability (see Chapter 14). Consideration should be given to designing the form that is to be completed in such a way as to remove the need to continually repeat certain information on every entry, for example date, location and units of measurement, or product, that may in some circumstances only need to be stated once on the form, whether completed by hand or electronically. Traceability of raw materials and packaging materials to their initial inspection records is critical within the QMS/FSMS/FIMS to ensure that food safety, integrity and quality criteria have been met. These records will form part of the audit trail in the event of a product withdrawal or recall (see Chapters 11 and 27).

In-Process/
Production and
Finished Product
Inspection Records

13.6   It is vital that accurate records be maintained of all relevant production data, including actual processing details and performance, and end product quality. These records should be readily available for a predetermined time after the end of the expected shelf life of the product (see 13.10).

The manner in which these records are kept will depend on the requirements and size of the operation and can range from individual worksheets and diaries to full-scale computer systems. Any mandatory requirements, such as those for UK weights and measures legislation, must be implemented (see 13.10).

13.7   The use of clear concise notations should be ensured and all records made in non-fade permanent ink. The use of Tippex-like materials on written documents should not be permitted; any alteration of an entry should be done by strike-through, accompanied by the corrected entry and initials of the person making the correction. This function should also be available on electronic records. Any amendments that require formal authorisation, for example by the quality control manager, should be appropriately and clearly authorised. Wherever possible, all records should declare the units of measurement and the physical conditions of these measurements. The use of the British Standards Institution (BSI) notation is recommended. Records should, wherever possible, bear details of the target and actual values recorded.

13.8   Personnel completing records should have sufficient training on how to complete the form, and the effectiveness of this training should be assessed routinely. Particular attention should be paid to ensuring that personnel understand the need to fully complete the record. Personnel must be aware that a blank box on the record implies that the inspection or test has not been undertaken, that is, there is no evidence to support it has been undertaken. The quality control manager should ensure that all quality control personnel show consistency in how they complete the records and also assess the use of quantitative rather than semi-quantitative or qualitative measures of quality criteria. The use of scoring systems or colour-coding systems (such as a traffic-light system) for records should be assessed, especially where criteria are being used to provide data to monitor performance against food safety and quality objectives. Terms such as 'good', 'nice' and 'ok' are qualitative and may represent in practice different standards of quality as assessed by individual staff. Appropriate training associated with developing quantitative measures to assess product quality should drive greater consistency between individuals, shifts and processing units (see Chapter 17).

Training of production and quality control personnel should emphasise the importance of completing all records at the time the activity is undertaken for which they are recording the data. Personnel must not fill in a succession of ticks and/or initials at some later convenient time. Records should be designed where possible for actual numbers, code letters or words to be written rather than just a tick or cross. Repetitive ticking of a record can lead to checklist fatigue for the personnel completing the record, and this must be avoided at all costs. It is very important that the individual initials each set of information that (s)/he completes on a record so that in the event of non-compliance, there is traceability on the records to the particular individual who undertook the task. This will also provide an audit trail to associated training records in order to confirm the training that individual has received and whether it has been appropriate and remains current, that is, that they are competent to undertake the tasks for which they have signed or completed the record. Personnel such as the quality control manager or designate, who are required to countersign or 'sign off' quality records, should formally identify whether the countersignature identifies the form has been received and is ready for filing or if the countersignature demonstrates that the information on the record has been independently verified and signed off as accurate and complete.

Master Forms   13.9   The term 'master form' here relates to both paper (hard) forms and electronic systems of data collection. To assist in the design and construction of master forms and the data they then contain, whether in paper or electronic format, the following checklist may be of assistance for the quality control manager:

(a) Determine the person(s) responsible for developing, designing and setting up the necessary master form(s) and determine the competent personnel who will be undertaking the checks and then completing the record(s). Suitable training must be given to ensure adequate understanding of the role required of inspection personnel and the significance of identifying system or product non-conformance and the action to be taken.

(b) Ensure that results may be recorded easily and there is sufficient space on the master form to adequately record the results of inspections.

(c) Ensure that results and data can be transmitted promptly to those personnel who require it.

(d) Arrange that results and data can be transmitted to the proper and appropriate personnel and that the recipients are in a position to take action as and when necessary.

(e) Ensure there is sufficient space to record preventive and corrective actions taken as a result of inspection.

(f) Ensure that re-inspection activities, if required, can be appropriately recorded.

(g) Control charts may frequently be used to replace tabulated forms with the major advantage that the situation may be noted at a glance and not lost within a mass of numerical values. Where appropriate, data logging for automatic control systems should be used (see Chapter 39).

(h) Wherever possible case markings, batch numbers, delivery vehicle and container numbers, along with container seal numbers, are all items of information that should be recorded when examining incoming and outgoing goods. All documents, samples and related data should be clearly dated, if necessary with the time, day, month, week number and year. The use of a sequential or consecutive numbering system is to be commended.

(i) Ensure that if samples of packaging and labels are going to be attached to the paper forms there is an indication of where they should be attached, the time they were attached, that is, whether quality assurance or quality control processes are adopted, and a mechanism for ensuring that such labels are formally signed off. If the packaging or label information is scanned into an online system there also must be an indication of where this information is stored and how such processes are formally approved and details retained on the electronic system of who formally signed off the packaging or labels as suitable for use.

(j) Ensure that master forms are routinely reviewed to ensure that the design of the master form does not impinge on the ease of completion.

Control of Records Procedure   13.10   Not all records may need to be kept after the production period has ended. The QMS/FSMS/FIMS should not require the retention of excessive records and should ensure that a protocol is set up and followed as to how long individual records should be kept, that is, for a defined time period. Special attention should be given to the shelf life of the product and any legal requirements, including the records and their value in the provision of evidence of due diligence. A controlled records list involving an ongoing and continuously monitored system for purging the files of unwanted old data should be adopted. The control of records procedures should identify how records pertaining to the QMS/FSMS/FIMS and 'one-up' 'one-down' traceability are collated, reviewed and verified, and maintained, including storage and retrieval. In the event of a product recall, customer complaint or other incident, it may be required to retrieve records in a short period of time, for example less than 2 hours, less than 4 hours or within a day, and this should be stated within the procedure and tested a minimum of annually as part of the product recall testing process. All records and data should be carefully examined by a responsible person before filing and at the same time screened for any irrelevant data, which should be removed. The retention time for documents and records should be determined by assessing the shelf life of the product or the shelf life of the customer's product if manufacturing ingredients or the product is suitable for freezing by the consumer. Any alterations to records should be authorised and dated, and justification for the alteration should be recorded (see 13.7).

Document Approval, Amendment and Control Procedure   13.11   A working method should be established and documented to ensure a document approval procedure is in place prior to issue, and only those personnel with the appropriate level of authority have approved the content and authorised the release of the document. This procedure should be in place for both paper-based and electronic-only systems.

13.12 A further working method should be established and documented to ensure that all documents, specifications and so on that form part of the formal management system are updated, that any current amendments are maintained and superseded documents are removed from the system, but held on file for a pre-determined period of time (see 13.10). The reason for any changes or amendments should be recorded so that there is a suitable audit trail in place. An adequate indexing and recall system, including records of amendments, is essential. The use of systems such as master indexes or master logs is recommended as a means to identify the current version of individual documents and master forms which are currently in use. The working method should also include the method of document identification and the system for replacing existing documents when they have been superseded, especially where there are multiple copies in circulation. Documents should remain legible. Care should be taken when photocopying master copies of paper documents to ensure that neither the quality of print is lost over time, nor a portion of the document is 'lost' due to incomplete copying, for example the document control information on the bottom of the document is not copied. Attention should be paid to the number of copies of a master copy required so that the numbers reproduced will be used in a timely fashion and will be adequately controlled. This will minimise the potential for obsolete master forms being used to record information when the form has been updated.

13.13 Internal audits should be routinely undertaken to ensure that the appropriate version of documents and master forms are being used (see 11.9).

13.14 Batch and laboratory samples, when stored, should be treated very much in the same way as documentation, records and data, but special emphasis must be placed on recording the physical conditions of storage, for example temperature, humidity and light intensity. All labels should be firmly affixed to the container and be checked to ensure that they are durable for the period of expected life. The date on which they may be discarded should be clearly defined.

13.15 Documents received and retained in electronic format should be adequately controlled to ensure they are not inadvertently deleted or amended. Provision should be made within the procedures to ensure that there is a back-up procedure in place for all electronic data to minimise the risk of losing data. Fire risk must also be assessed, and contingency plans need to be in place with regard to paper-only systems such as a fireproof safe for master copies/key documents or Cloud storage etc. (see Chapter 39).

14 PRODUCT IDENTIFICATION AND TRACEABILITY

Principle

It may not necessarily be enough to assume that the description of a consignment of a raw material on the packages or on the corresponding invoice is accurate. Where the identity is not absolutely obvious beyond question, the identity of each consignment of raw material should be checked to verify that it is what it purports to be. Product identification and traceability systems must be able to trace materials from source to finished product and also from product to source. The system should be routinely tested to demonstrate that it is implemented and effective. Product identification and traceability procedures are a key prerequisite to ensuring good manufacturing practice (GMP).

Product Identification

14.1 Examples of raw materials whose identity may be obvious beyond question are many fruits and vegetables (although where a particular variety or cultivar is specified, the manufacturer receiving a consignment will need to determine that it is of the cultivar specified).

14.2 Major examples illustrating the dangers of simply assuming that raw materials are what they are supposed to be are common. The inadvertent use, by some manufacturers, of what subsequently was revealed to be unfit raw materials can lead to contaminated food batches that are then subject to recall or, where an animal feed manufacturing operation is involved, the slaughter of animals in the supply chain who have been potentially fed contaminated feed, for example the problems that have occurred in the past with dioxin-contaminated feed.

14.3 In evaluating a supplier's premises and operations, attention should be paid to, and account taken of, the safeguards operated by the supplier against mistaken identity of materials supplied to them and then subsequently materials leaving their facilities.

14.4 In some instances, checking a raw material delivery for compliance with its specification will also establish its identity. Where this is not the case, consideration should be given to building into the inspection schedule a check to verify the identity of the material received. It would be a mistake to assume that because an ingredient may be a clearly defined chemical entity, supplied by a reputable supplier who gives assurances as to compliance with legal purity requirements, that its identity may be taken for granted. Chemical manufacturers who manufacture food additives often also manufacture other chemical substances for non-food use and thus are not food grade, and mistakes or mislabelling or ill-intent could occur. Due to the seriousness of the potential food safety hazard or food integrity threat (see 5.1) involved in such an occurrence, it is all the more important that each

Food & Drink – Good Manufacturing Practice: A Guide to its Responsible Management, Seventh Edition.
The Institute of Food Science & Technology Trust Fund.
© 2018 John Wiley & Sons Ltd. Published 2018 by John Wiley & Sons Ltd.

consignment of an additive or ingredient should also be checked to verify its identity and that it is food grade and it is what it purports to be. In the instance that fruits and vegetables are imported, then on receipt it is important to assess that the designated country of origin (COO) is recorded and it is verified that it is indeed correct. Where imported food is processed in establishments approved through either legislative means or third-party certification, it is important that product identity and traceability to a given processing unit are established on physical labelling and/or the associated documentation and verified by staff at the manufacturing unit on receipt. If it is not possible to verify this at receipt verification must occur before use, for example if container consignments cannot be broken down on receipt for verification of all pallets, boxes or other net units of supply contained within the batch the identity of the units must be determined at the start of production activities, when it is practicable to check and verify such information.

Traceable Resource Units   14.5   It is important for the manufacturer to have a clear communication strategy for the terminology used to determine traceability and describe how traceability is embedded within their systems and procedures. The term traceable resource unit (TRU) is often used. This has been described as a unique batch of material or product that, using a specific set of traceability characteristics, is readily distinguishable from other batches of material. A batch is therefore an identifiable amount of product or material that has undergone a given treatment or process at a given time in a given location. ISO 20005:2007 defines a lot as 'a set of units of a product which has been produced and/or processed or packaged under similar circumstances'. Alternatively, vocabulary is used in the manufacturing sector such as *lot* (a group of items defined by the manufacturer; see also 14.9), *trade unit* (a quantity defined at its lowest repeatable unique unit, e.g. jar, box, pack, bag) and *logistic unit* (a quantity defined at its lowest repeatable unique unit for storage or transportation, e.g. tray, box, pallet or container).

14.6   The challenge for the food manufacturer is that a given TRU is constantly changing in terms of physical and traceability characteristics within the manufacturing process. The manufacturing activities employed, including mixing, addition, separation, splitting, aggregation, joining, storing, transfer, rejection and reassembly, mean that continuously new TRUs are created. Many manufacturing operations are not in themselves discrete, i.e. individual batches produced one at a time that can be identified by time and location (e.g. bin or tank number or time of cleaning between one day of production and the next), instead they are continuous and cannot be differentiated. An example of continuous batches is the ongoing refilling of a bulk bin with a raw material, e.g. milling wheat on receipt in a bulk bin in a flour mill or a bulk ingredient silo in a manufacturing unit. Thus each manufacturer needs to be very clear as to which of their processes are discrete and which are continuous, and as a result how they are going to develop a product identification and traceability procedure that is appropriate, implemented, monitored and verified on a regular basis to ensure its efficacy.

14.7   Traceability has two aspects: first, the ability to *trace* a product back to the start of the production process and, second, to *track* a product from a given point in the manufacturing process to despatch, then subsequent stages and ultimately to the consumer. This means that together with the physical movement of material from stage to stage in the process there also needs to be an associated information system. The information system can create transparency within the manufacturing process and capture a number of details about the product. These criteria include material and product history, the time the material spends at different stages of the supply chain and in whose ownership, who was involved with making the product, how the product has been stored, which is especially important if temperature control is critical (e.g. the cool-chain), the diverging of product flows if materials are used in different products and the destination of by-products, waste (e.g. if it is used in animal feed) and details of provenance such as country of origin, method of production on farm (such as organic, conventional production), geographic origin etc.

Traceability   14.8   The **EU Regulation No. 178/2002** defines traceability as the ability to trace and follow a food, feed, food-producing animal or substance intended to be, or expected to be, incorporated into a food or feed through all stages of production, processing and distribution. The legislation requires that an item can be traceable 'one-step forward and one step back', sometimes called business-to-business (B2B) traceability. Market requirements may mean that traceability requirements are more in-depth, requiring businesses to trace from field to fork. This means that not only is there physical movement of product from step to step in the supply chain, there is also a data trail from the start to the end of the chain. Manufacturers need to determine what their TRU is in terms of a batch or a lot. However, with reassembly and mixing during the manufacturing process, as previously described, the definition of the batch can change and new batches are created that need to be linked to the previous steps in the chain. As a result, whilst regulation 178/2002 outlines the minimum legal requirement for traceability, market requirements are increasing the need for technology to capture wider amounts of data (see Chapter 16).

Lot Marking   14.9   The **Food (Lot Marking) Regulations 1996** (as amended) set out
Regulations   the lot marking requirements to be applied to all foodstuffs sold for consumption (unless specifically exempted), including wines and spirits. They implement **Council Directive 89/396/EEC** on indications and marks identifying the lot to which a foodstuff belongs as amended by **91/238/EEC**[1] and **92/11/EEC**. The directive

---

[1]http://eur-lex.europa.eu/legal-content/EN/TXT/?uri=CELEX%3A31991L0238.

establishes a framework for a common lot (or batch) identification system throughout the EU so that a product recall can be implemented if required. It is for the manufacturer or other appropriate business in the supply chain to determine the size of a lot. It is important that the size of the lot reflects the practicality of a potential product recall. The **Food (Lot Marking) Regulations 1996** (as amended) defined a lot as 'a batch of sales units of food produced, manufactured or packaged under similar conditions' and lot marking indication as 'an indication which allows identification of the lot to which the sales unit of food belongs'.

Genetically Modified Organism Regulations

14.10 **EC Regulation No. 1830/2003** concerning the traceability and labelling of genetically modified organisms and the traceability of food and feed products produced from genetically modified organisms (GMOs) requires that 'traceability requirements for GMOs should facilitate both the withdrawal of products where unforeseen adverse effects on human health, animal health or the environment, including ecosystems, are established, and the targeting of monitoring to examine potential effects on, in particular, the environment. Traceability should also facilitate the implementation of risk management measures in accordance with the precautionary principle'.[2]

Traceability requirements for food and feed produced from GMOs must be established to ensure accurate labelling in accordance with the requirements of **EC Regulation No. 1829/2003** on genetically modified food and feed to ensure accurate information is available to operators and consumers and to enable control and verification of labelling claims. In the UK, the provisions from these statutes have been incorporated into the **Animal Feed (Composition, Marketing and Use) (England) Regulations 2015** and equivalent legislation.

Provenance, Identity Claims and Assurance Status

14.11 Where claims are made on the finished product, it is important that the ingredients or the product itself in reality comes from the said designated provenance, or has the designated identity or assurance status in practice, for example Fairtrade, organic, or designated COO. The manufacturing organisation must have a suitable identification and traceability procedure in place to demonstrate the status of raw materials, in-process material and finished product, especially with regard to such claims. Records should be maintained to demonstrate that the materials sourced and the finished product comply with the legal requirements associated with such claims. This is discussed in more detail in Chapter 15.

Traceability Procedure

14.12 BS EN ISO 22000:2005 *Food safety management systems – requirements for any organisation in the food chain* states that a traceability system should establish and enable the identification of product lots and their relation to batches of raw materials,

[2]http://eur-lex.europa.eu/LexUriServ/LexUriServ.do?uri=OJ:L:2003:268:0024:0028:EN:PDF.

processing and delivery records. It is important to consider the scope of the product identification and traceability system, and the associated data requirement and whether the scope of the traceability system includes food safety and legislation, food quality and food integrity within its boundary of operation.

Four types of traceability systems have been defined:

- track-and-trace based integrity systems concerned with identity preservation through each step in the process;
- volume-based integrity systems concerned with the ability to track by weight or volume where a given TRU has gone. These types of traceability systems are often called mass-balance systems;
- separation-based integrity systems concerned with segregation and the ability to isolate a given TRU from others; and
- certificate-based integrity systems concerned with how identity and traceability can be assured by the data and documentation that accompanies the TRU. These are sometimes called book and claim systems.[3]

It is important for the manufacturer to be clear as to the type of traceability system they have adopted to ensure food safety, legality and quality have been maintained. A documented procedure should be developed and implemented to ensure that, throughout all stages of an organisation's operations, raw materials and in-process items and finished products are identified and traceable. Packaging and processing aids should also be traceable to an individual delivery and/or batch. The quality control manager is responsible for the implementation and validation of this procedure. Guidance on the general principles and basic requirements for traceability system design and implementation is contained in ISO 22005:2007.

14.13   In practical terms, for traceability to be effective, the responsibility of each organisation is to ensure the 'link' in the food supply chain is not broken. This requires an organisation to satisfy itself that its supplier of a food material has carried out checks to ensure that the documentation provided accurately describes the nature and substance of that material and its status. The organisation should retain that documentation and, following the manufacturing process, issue documentation/labelling to the organisation to which they in turn sell the food product which can link back to all previous processes and stages that the associated TRU has gone through. Where reworking/repacking operations are being performed, traceability must be maintained through quality control documentation.

14.14   The product identification and traceability procedure should include both internal traceability, i.e. of products and materials

[3]Mol, A.P.J. and Oosterveer, P. (2015). Certification of markets, markets of certificates: Tracing sustainability in Global Agro-Food Value Chains. *Sustainability, 7*(9), 12258–12278.

within the control of the manufacturer and external traceability that covers tracking and tracing at all stages of the supply chain. The quality control manager or designate should ensure that the product identification and traceability procedure is tested at prescribed intervals. These tests should interface with the crisis management, complaints and product recall proctols, and the testing of traceability should include a variety of potential scenarios (see Chapter 27). This traceability test should include a mass balance or quantity test (see above), i.e. that the volume/weight of the product in the designated TRU can be fully traced/tracked from intake to despatch or vice versa. This requires any reworked, regraded or waste product to be fully recorded within manufacturing records. As part of the management review and internal audit process each manufacturing unit should also undertake a formal assessment to determine the level of traceability achieved in practice and whether in the event of a product recall further batches should be automatically recalled. This assessment should also include a determination of the timescales that are acceptable for the traceability audit to be completed, for example what documents should be available within 2 hours, 4 hours, 8 hours and so on.

15 PROVENANCE AND AUTHENTICITY

Principle

The term 'food provenance' relates to not only the geographic elements of where the ingredients and the final food are grown, processed and finally manufactured, but also how that food is produced and whether the methods of production and processes employed comply with certain standards and protocols. Authentic products are those that demonstrate a given connection to a recipe, location or social characteristic. Authenticity is the innate quality of being authentic, genuine and of undisputed origin. It is essential for the food manufacturer to design, validate, implement and verify that adequate and appropriate systems are in place to ensure the provenance and authenticity of food ingredients and food products.

Background

15.1 Provenance as an innate characteristic relates to where and how ingredients are grown, caught and reared, the types of farming system employed (e.g. conventional, intensive, extensive, sustainable, organic, housed livestock, free-range, genetically modified (GM), farmed or wild-caught fish and so forth), the types of production or processing (e.g. traditional, home-cured, oak-matured etc.), the standards of people welfare said to have been complied with in the supply chain (e.g. Fairtrade, Ethical Trading Initiative Basecode, SEDEX) and relates specifically to the manufacturer's geographic location and situation and the portfolio of materials used and products manufactured (e.g. local, seasonal, breed of livestock, specific crop variety etc.).

15.2 Food provenance can relate to a location, e.g. national or regional, down to a specific farm or field. Specific product claims and labelling will reflect this, such as country of origin (COO) labelling. Regulatory requirements with regards to COO labelling vary from country to country. The **Food Information to Consumers' (FIC) Regulation** implements **EU [European Union] Regulation No. 1169/2011** on the provision of food information to consumers. Regulation No. 1169/2011 identifies the origin of a food as being either its COO (see Articles 23 to 26 of Regulation (EEC) No. 2913/92) or its 'place of provenance'. The term COO is defined in EU law by Articles 23 to 26 of EU Regulation No. 2913/92 establishing the Community Customs Code, being either:

- *Goods originating in a country shall be those wholly obtained or produced in that country,* for example vegetable products harvested therein, live animals born and raised therein; and
- *Goods whose production involved more than one country shall be deemed to originate in the country where they underwent their last, substantial, economically justified processing or working in an undertaking equipped for that purpose and resulting in the manufacture of a new product or representing an important stage of manufacture.*

Food & Drink – Good Manufacturing Practice: A Guide to its Responsible Management, Seventh Edition.
The Institute of Food Science & Technology Trust Fund.
© 2018 John Wiley & Sons Ltd. Published 2018 by John Wiley & Sons Ltd.

Article 2(2)g of Regulation No. 1169/2011 states that a place of provenance is 'a place where food is indicated to come from, and that is not a COO… [however] the name, business name or address of the food business operator on the label shall not constitute an indication of the COO or place of provenance of food within the meaning of the Regulation.'

15.3   The EU FIC Regulation requires origin indication for fresh, chilled and frozen meat derived from sheep, goats, poultry and pigs. Beef also requires this labelling but through separate legislation. If animals are born, reared and slaughtered in the same country then 'reared in' and 'slaughtered in' statements may be replaced by a single origin declaration. Where meat from different locations is packed together from separate member states within Europe and/or third countries all countries must be listed. Where meat is imported from third countries the term 'non-EU' may be used.

15.4   The FIC legislation also requires that when the origin of a whole product is voluntarily identified that the origin of its primary ingredient, where it is different in terms of origin, must also be given, e.g. 'Made in the UK with New Zealand lamb'. A primary ingredient is one that represents more than 50% of that food or which is usually associated with the name of the food by the consumer and for which in most cases a quantitative indication is required (see Article 2(2) (q) of EU Regulation No. 1169/2011).

15.5   Usually, unless there is specific legislation that states otherwise, COO or place of provenance identification is only mandatory if its absence would mean that the consumer might be misled as to the true origin of the manufactured/processed food. Origin labelling is mandatory for certain products in the EU under specific product legislation rather than the aforementioned FIC legislation. These products include fish, virgin and extra-virgin olive oils, unprocessed beef and beef products, prepacked poultry meat from third countries, honey, fruit and vegetables.

Traditional and Regional Food   15.6   Traditional and regional food is prepared and made in a defined way, often using local ingredients. In the EU there are three indications of provenance for regional products with a given traditional or speciality heritage through the so-called EU Protected Food Name Scheme (EUPFN). The EUPFN is covered by **EU Regulation No. 1151/2012**[1] and was introduced in Europe via **EU Regulations Nos. 2081/92** and **2082/92** of 14 July 1992 defining the standards for a designation under different collective trademarks in 1993. The three categories are:

•   *Protected Geographic Indication (PGI)*: This designates products which must be produced, processed or prepared within the geographical area and have a reputation, features or

[1] https://ec.europa.eu/agriculture/quality/schemes_en.

certain qualities attributable to that area. At least one of the stages of preparation or processing or production must take place in that area. Examples are Scotch Beef, Arbroath Smokie, Welsh Lamb and Cornish Pastie.

- *Traditional Speciality Guaranteed (TSG)*: This identifies products which are traditional, e.g. a recipe or production method, or have customary names and have a set of features which distinguish them from other similar products. These features are not due to the geographical area the product is produced in nor entirely based on technical advances in the method of production.
- *Protected Designation of Origin (PDO)*: This mark identifies agricultural products and foods which are produced, processed and prepared in the same geographic area, e.g. Shetland Lamb and Anglesey Sea Salt.

15.7 Foods that are labelled under the EUPFN scheme could be vulnerable to food integrity threats such as imitation or mislabelling, especially where there is a price differential between the PGI products and other seemingly identical products that do not carry that status. Manufacturers that produce or use these products as ingredients need to be aware of this threat to integrity and develop a countermeasures programme through their food integrity management system or FIMS (see Chapters 5 to 7).

15.8 Where specific claims are made with regard to the manufactured product it is important that there is full traceability of the ingredients or the product itself, i.e. it may come from a designated provenance, or has a designated identity or assurance status, e.g. Fairtrade, organic, COO. The organisation must have a suitable identification and traceability procedure in place to demonstrate the status of raw materials, in-process material and finished product (see Chapter 14). Records should be maintained to demonstrate that the finished product complies with the legal requirements associated with such claims.

15.9 The terms 'authentic' and 'authenticity' are not as easy to define. Authenticity can relate to both intrinsic and extrinsic quality attributes (Chapter 2). Other terms have been used, such as *iconic attributes* or characteristics, i.e. those that relate to an object or event, and *indexical attributes*, which are characteristics that relate to facts or traditions. Often these attributes are distilled for consumers into terms such as traditional or artisan. It is important that such terms are suitably underpinned by records held by the manufacturer.

16 ELECTRONIC IDENTIFICATION AND DIGITAL TRACEABILITY TECHNIQUES

Principle

The use of electronic identification and digital traceability techniques either within the manufacturing unit or across the wider supply chain is well established. This chapter seeks to give an overview of the types of electronic identification methods used and their value with regard to demonstrating control of food safety, legality, quality and integrity criteria.

Background

16.1 The procedures employed by the manufacturer to ensure effective control of product identification and traceability, provenance and authenticity have been considered in Chapters 14 and 15. It is important for the manufacturer, when considering the use of electronic identification and digital traceability techniques, to first consider the attributes that they seek to monitor. These can be considered in terms of:

- *breadth*: the number of attributes, or the quantity of information connected to each traceable resource unit (TRU) (see 14.5);
- *depth*: how far upstream or downstream from the manufacturer the TRU needs to be monitored;
- *precision*: the degree of accuracy in pinpointing a particular TRU's location, movement or characteristics; and
- *access*: the speed or timeliness with which tracking and tracing information needs to be communicated to the manufacturer.[1]

16.2 Whilst **EU Regulation No. 178/2002** outlines the statutory requirement for traceability within the European Union (EU), and the need for one step forward and one step backward traceability (see 14.8), market requirements to be able to track and trace from field to fork and concern over maintaining integrity in supply chains means there is an increasing need for manufacturers to use technology to capture wider amounts of data and make sure it is accessible in the timescale required for routine monitoring, verification or in the event of a product withdrawal or recall. Technologies being used to develop appropriate data traceability systems are what are collectively described as automatic identification data capture (AIDC) systems. For a given TRU, AIDC systems link together data from the various electronic sources about the individual items into a cohesive history that can then be shared by supply chain partners. The result is that the speed of data flow can be increased, costs can be reduced and it supports

---

[1] Dabbene, F., Gay, P. and Tortia, C. (2014) Traceability issues in food supply chain management: A review. *Biosystems Engineering*, **120**, 65–80.

Food & Drink – Good Manufacturing Practice: A Guide to its Responsible Management, Seventh Edition.
The Institute of Food Science & Technology Trust Fund.
© 2018 John Wiley & Sons Ltd. Published 2018 by John Wiley & Sons Ltd.

greater data security in a global supply chain. As a result, through the use of information management systems and real-time technology such as electronic data interchange (EDI) or extensible markup language (XML), data-based traceability systems can be developed by the food manufacturer. There are different levels of TRU, e.g. trade unit, logistics unit and lot (see 14.5), and supply chain partners may use different technology to collect data at each level that when combined forms an overarching traceability system that links through individual product packs to the logistic unit at box or pallet level through to the overall physical lot, i.e. the definable batch of product. This is important to consider as the traceability process has to work through the entire logistics and distribution chain to when a product is purchased on the shelf in a retail environment. Data can also be captured with regard to expiry dates within the AIDC system so this has a value not only with stock control, but also as technology develops in the home setting too.

Data can be exchanged between businesses using EDI, enabling multiple electronic messages such as logistics information to transfer between the originator and the recipient. Data is captured via 1D (series of parallel bars) stacked 1D (multiple bar codes above each other), and 2D (such as quick response (QR) bar codes via scanning of labels on trays, cases and pallets. QR codes contain a variety of information, including batch numbers, duration dates and weights or volumes and are of benefit in addressing the risk of counterfeiting of raw materials and finished products. Optical character recognition (OCR), and radio frequency identification (RFID) tags are also of value when developing electronic identification and digital traceability systems.

16.3   RFID tags require two physical elements in order to operate effectively: a receiver (sometimes called reader or interrogator) and a transponder (sometimes simply called a label or tag). The use of RFID tags means that the product can be identified and tracked using radio waves. RFID is of value during manufacturing processes to identify continuous TRU as well as discrete batches. There are challenges to using RFID in a manufacturing environment where it might be hot, cold, wet and this can lead to physical limitations with its use. The nature of the food item in terms of composition, moisture content, shape and size will also influence the effectiveness of the RFID system, for example foods with a lower water activity will have less influence on the operational viability of the tag. Food packaging materials such as glass and metal can also have a negative effect so the manufacturer must ensure that they consider this during the traceability system development process. It may be necessary, for example, for the product to be placed in secondary packaging to which the RFID tag can then be attached. Thus tags can be applied to a container, tray or pallet rather than individual products.

16.4   Other challenges with developing AIDC systems are the lack of uniformity with different equipment and format standards, and this can limit how information is shared. The level of granularity of the system is important. Granularity describes the depth and detail of the information required. A low granularity system works at the summary level whilst a high granularity system works at the individual transaction, i.e. at a minute level. The level of granularity therefore will determine whether the traceability system is operating at item, case or pallet or container level. The volume of data being managed, cost, the skillset required to operate the system and the need for stationary or mobile sensors and readers all influence how the system is ultimately designed and implemented in the manufacturing environment. Global positioning systems (GPS) together with RFID tags and sensors can be used to control the chill chain and on-farm electronic identification (EID) of animals via tags is an example of the use of low-frequency RFID. Therefore the manufacturer needs to consider how the AIDC system elements are used internally within the manufacturing unit and their compatibility with those system elements used externally across the whole chain from farm through to retail or food service.

16.5   Electronic technologies, including geographic information systems (GIS) and remote sensing (RS), Bluetooth, WiFi, WiMax, WiBro, Zigbee, Ultra-Wide Band and web services, mean that data management systems, frameworks and/or architectures can be implemented across the supply chain.

16.6   The use of electronic traceability and identification techniques aids a number of the procedures and protocols outlined as being key elements of an effective good manufacturing practice system. Examples include *barcodes* being used in inventory control, stock allocation, stock and batch movements, theft protection and anti-counterfeiting, *commercial traceability software* used for tracking and tracing products and food safety and quality assurance information through stages of the manufacturing process and the wider supply chain, *EID*, which supports batch traceability and management, and *RFID*, which is used for information sharing, temperature-time control, livestock management, theft-prevention, electronic payment, automated production systems, navigation management, inventory management and promotions management.

16.7   Electronic sensors can be combined with electronic traceability and digital identification techniques to give further information about product and process characteristics. Examples of these sensors include biosensors that can identify analytes, microorganisms and proteins in food, chemical sensors that can monitor food quality and packaging integrity, nose systems that provide information on ripening, fermentation or spoilage, and traditional sensors that can monitor temperature, humidity and so on.

16.8   Blockchain is another technological solution cited as being of value in tracking and tracing food products, especially in the event of a product recall. The advantage of such technology is the speed of data capture and analysis, the development of data platforms and the accessibility of the data to multiple stakeholders in the event of a food safety, quality or food integrity challenge. However, this advantage rests on the effective use of such systems by people across the whole food chain, not just in manufacturing. These individuals require the designate skills, literacy and training to use the technology. The efficacy of data input is the limiting factor throughout what are often global chains, especially where some production stages have limited technological access and conditions can be adverse, e.g. scanning of livestock ear tags in fields in what can be poor weather conditions. However, the technology is developing fast and such challenges will potentially be overcome in the short to medium term.

17 PERSONNEL, RESPONSIBILITIES AND TRAINING

Principle   *There should be sufficient personnel at all levels with the ability, training, experience and, where necessary, the professional and technical qualifications appropriate to the tasks assigned to them and commensurate with the size and type of the food business. Their duties and responsibilities should be clearly explained and recorded as job descriptions with associated organisation charts or by other suitable means. Training should cover not only key tasks, but also good manufacturing practice (GMP) generally, and the importance of, and factors involved in, personal hygiene. Personal hygiene protocols are a key prerequisite to ensure GMP.*

General   17.1   Manufacturing companies vary widely in character, size and structure, whether they are independent or part of a group, and, in the latter case, the extent to which the character of their operations is subject to central control. There are also wide differences in terminology and job titles in different companies. These circumstances make it impossible to adopt rigid generalisations about the titles of key personnel but the principles outlined in this chapter are of importance, and the terms 'production manager', 'quality control manager' and 'purchasing manager' are used here for the sake of convenience, but may equally well be read as 'general manager', 'operations manager', 'technical manager', 'quality assurance manager', 'procurement manager' rather than the job title, as the responsibilities of the departments concerned.

17.2   Two key personnel are the production manager and the quality control manager, who should be two different people, neither of whom is responsible to the other, but who both have a responsibility to collaborate in achieving product safety, legality and the required product quality. A third key person is the purchasing manager, who has a special responsibility for ensuring that raw materials/ingredients are purchased in compliance with specifications, and for collaborating to that end with the quality control manager. A supplier quality assurance protocol that includes supplier performance monitoring should be developed for the manufacturing company following collaboration between the purchasing manager and the quality control manager (see Chapter 23).

17.3   Persons in responsible positions should have sufficient authority to discharge their responsibilities, which should be clearly defined (see 17.8). In particular, the quality control manager should be able to carry out his/her functions impartially and in accordance with the Institute of Food Science & Technology (IFST) Code of Professional Conduct (http://www.ifst.org).

Food & Drink – Good Manufacturing Practice: A Guide to its Responsible Management, Seventh Edition.
The Institute of Food Science & Technology Trust Fund.
© 2018 John Wiley & Sons Ltd. Published 2018 by John Wiley & Sons Ltd.

17.4   The quality control manager should possess appropriate professional and scientific/technological qualifications, and experience in food safety, food legality and food quality control. She/he should also possess interpersonal skills so that she/he can effectively communicate food safety, legality, integrity and quality criteria at all levels within the organisation.

17.5   Key personnel within the manufacturing organisation should be provided with an adequate number and calibre of supporting staff. A formal human resource appraisal should be undertaken with consideration given to the requirements of developing, validating, implementing, monitoring, revalidating and verifying the hazard analysis critical control point (HACCP) plan(s), the food safety management system (FSMS), the food integrity management system (FIMS) and the quality management system (QMS) and associated quality plans. These formalised systems may be separate or may be integrated into one formal documented system. The human resource appraisal review should include the identification of the human resources required to implement these system(s) consistently and effectively. A further review should be undertaken, minimally each year or more frequently when operational circumstances change that impact on the resource requirement needed. These reviews and any associated actions should be documented.

17.6   Key personnel within the manufacturing organisation should be direct employees of the company. Persons contracted, but not employed directly by the company, may be retained to provide advice or to carry out special projects, but should not be regarded as filling key personnel positions. During the approval process for contract personnel, it is important to determine the level of qualifications, current validity of membership of professional bodies and the status of continuous professional development (CPD), and, wherever possible, seek references from previous clients.

17.7   Persons should be designated to take up the duties of key personnel during the absence of the latter. This should be formally documented, and the organisation must be able to demonstrate that deputies are competent to undertake the role when required.

17.8   The ways in which those responsibilities that can influence product safety, legality, integrity and quality and the authority to discharge them are distributed among key personnel may vary with different manufacturers. What is essential in all organisational arrangements is that responsibilities should be clearly defined and allocated and areas of authority delineated. These arrangements should be properly understood and respected by all concerned. This may be documented in job descriptions/specifications, organisation charts or alternative documents.

17.9 The quality control manager should have the authority to establish, validate, implement and verify all food safety, legality, integrity and quality control procedures, to approve materials and products, and to withhold approval from any material or product that does not comply with the relevant specification. In some organisations, withholding of approval amounts to absolute rejection; in others, it amounts to provisional rejection subject to confirmation by the general management or the managing director; in yet others, it may be absolute rejection in certain defined serious cases, and provisional rejection in others. In exercising authority in the foregoing ways, the quality control manager, subject to the proper discharge of responsibilities, is under obligation to do everything possible to minimise and preferably prevent any disruption of production and/or distribution arrangements and schedules, and to provide prompt information and advice to production personnel to help maximise conformance with specification, and, in the case of non-conformance, to provide advice on rectification or reworking if possible. The quality control manager should, where required, identify whether the rejection is on the grounds of legislative, food safety or quality criteria to ensure suitable advice to production staff and to enable appropriate corrective action. The quality control manager has a leading role in defining preventive action where weaknesses in the QMS/FSMS/FIMS and/or compliance in practice with the QMS/FSMS/FIMS have been identified and non-conforming product has not yet been produced. The production manager should fully support such initiatives and enable preventive action to be implemented and effective.

17.10 The production manager should have responsibility for manufacturing resources, including processing areas, personnel, equipment, operations and records, and for using these resources to produce, in accordance with the master manufacturing instructions or equivalent, the programmed quantities of products conforming to the relevant quality specifications and within the budgeted cost. In addition, the production manager usually has some responsibilities that are shared, or exercised in liaison with, the quality control manager and other personnel. The production manager should be a member of the HACCP team and should participate in the development and acceptance of specifications, particularly the master manufacturing instructions and the finished product specifications, and in the training of production personnel. She/he should liaise with the quality control manager in ensuring an optimum manufacturing environment, in particular in relation to controlling and minimising extrinsic hazards (food safety hazards arising from the processing area, personnel, equipment and operations), hygiene and for quarantining and rectification of any non-conforming product. The production manager should not have authority to use any raw materials or intermediates that have not been approved for use, unless a concession has been formally agreed with the quality control manager or designate.

17.11 The quality control manager should have an advisory role in the choice of suppliers of raw materials, packaging and services (see Chapter 23).

*Recruitment
and Selection*

17.12 In the recruitment and selection of staff, due regard should be given to the required prerequisite skills and the potential suitability of the candidate for the task in hand. For example, persons of clean and tidy and methodical nature may be more likely to appreciate the principles required in the hygienic handling and storage of food and the implementation of the FSMS, FIMS and the QMS, than those who are not. Similarly, potential employees should be appropriately qualified for the requirements of the task as defined in the job description/specification.

Training

17.13 Training is necessary to enable personnel at all levels within the organisation to understand their responsibilities and improve their knowledge and skills. Effective training will also improve individual performance and reduce the level of supervision required. Training will only be effective when the provision of knowledge develops thorough understanding and is relevant to the individual's tasks and responsibilities. The implementation of the knowledge acquired by individuals requires management support, including appropriate opportunity to practise skills, ensure motivation and provide effective supervision. It should be noted that while a certificate of attendance at an internal/external training course is of value, that value is limited if it cannot be verified that the person(s) attending such training can effectively implement the skills and knowledge gained in practice in the manufacturing environment. Similarly signatures on a training record demonstrate training has taken place, but do not in themselves provide evidence of ongoing staff competence. Therefore the internal audit programme needs to include in its scope the verification of personnel competence at all levels of the organisation as well as the presence of training certificates, especially where their work activities can affect food safety, legality, integrity and quality.

Training records should be maintained and they should provide as a minimum details of the nature and content of the training undertaken, the date and duration of the training activity, the identity of the trainer and the signatures of the trainee to verify training has taken place and the trainer to confirm that the training was effective and the individual is competent. Individuals should only be signed off subsequent to the training course or activity when they have demonstrated this competence in all aspects of the job role to the satisfaction of the trainer or designate. Some job roles will require specific training, for example individuals involved in chemical handling, machine operating and pest control, and this must be documented and signed off again by the relevant person to demonstrate that the individual has attained a minimum level of competence. The initial and ongoing level of competence of the trainers should also be identified and documented by the manufacturing organisation in order to verify that they can provide appropriate training and knowledge transfer to staff. If agency staff are used, then their training and the competence of the agency trainers must be verified as suitable before they commence work. Additional site-specific training will be required in addition to familiarise agency staff with the specific aspects of the manufacturing process and the food safety, legislative and quality controls in place on the site. Full consideration should be given to the risk of using agency staff at critical control points (CCPs), critical quality points (CQPs) and critical legislative points in the manufacturing process with respect to their suitability and degree of competence for the role. Use of agency staff should form a component of the risk assessment process in ensuring the FIMS system is appropriate and effective.

17.14   All production and quality control personnel should be trained in the principles of GMP, and in the practice and underlying principles of the tasks assigned to them. Similarly, all other personnel (e.g. those concerned with maintenance or services or cleaning) whose duties take them into manufacturing areas or have a bearing on manufacturing activities should receive appropriate training. Records should be kept of the training of each individual. Where individuals are working at CCPs as defined in the HACCP plan and/or have responsibility for the effective implementation of prerequisite programmes (PRPs) or food integrity countermeasures, the manufacturing organisation must develop and implement training programmes to ensure that these individuals have appropriate levels of competence, skills and knowledge to support the effective adoption of the HACCP plan, QMS, FSMS and FIMS. Similarly, those individuals working at CQP and critical legislative points such as label printing, quantity control (weight, volume, count) and points of control of provenance, assured status and identity control should have appropriate training and supervision to ensure that they are competent, they understand their job role and its importance, and that legislative requirements are consistently met.

17.15   Training should be in accordance with programmes approved by the production manager and the quality control manager. Training should be given at recruitment (usually termed induction training) and repeated, augmented and revised as necessary. The degree of an individual's knowledge and understanding of the content and application in the workplace of the information contained in the induction training should be tested and verified before individuals can commence work unsupervised. It is essential to consider developing a 'buddy system' for new employees during induction so they can receive on-the-job training and support. In the event of non-conformance by staff with procedures and work instruction then retraining should be undertaken as required. Records of induction training should be retained for all personnel, and with returning seasonal workers induction training should be undertaken each season, with particular emphasis on changes in legislation, customer requirements and emerging food safety, legality, integrity and quality issues. Both in training itself and with regard to the need for personnel to be able to understand and follow written instructions and procedures, notices and so on, particular attention should be given to overcoming language, numeracy or literacy difficulties. Recommendations for the content of induction and supplementary hygiene training programmes are contained within the Industry Guides to Good Hygienic Practice.[1] These guides were initially developed in accordance with **Article 5 of the EC Directive on the Hygiene of Foodstuffs (93/43/EEC) Regulation (EC) No. 852/2004** and continue to apply after 1 January 2006, and continue to remain compatible with the objectives of the new legislation. Many older guides may not necessarily cover all of the revised requirements, and to date they have not all been reissued as revised guides. Ten guides have been subsequently published that address the following sectors: mail order food, wholesale distribution, sandwich bars and similar food outlets, flour milling, vending and dispensing, bottled water, sandwich manufacturing, whitefish processors, spirit drinks and retail. While not all the guides are in sectors defined as a manufacturing process, manufacturing organisations need to be aware of the requirements of subsequent sectors in the supply chain that they supply in order to ensure supply chain compliance.

17.16 Periodic assessments of the effectiveness of training programmes should be made, and checks should be carried out to confirm that designated procedures are being followed. Training programmes can be evaluated by a number of techniques, including:

- pre- and post-training observation and questioning to assess if knowledge has been gained, practices or standards have improved or attitudes have changed;
- post-training assessment to determine the individual's competence to produce safe food, which may involve written tests, observation or questioning, microbiological swabbing of hands and/or food contact surfaces or bacterial food sampling; and
- trend analysis of the frequency of customer complaints, quality incidents or level of waste or rework or the frequency of failures at CCPs and CQPs.

17.17 Training should be reinforced by adequate supervision, mentoring and support, regular performance reviews, refresher training at designated intervals, posters and notices, reviewing the results of routine microbiological monitoring and quality performance with personnel and recognising good practice. Refresher training

[1] These are available from The Stationery Office (http://www.tsoshop.co.uk).

is required where any element of the QMS, FIMS or FSMS changes, especially the induction programme, so that all staff are consistently trained to the same level. The time period as to when a person is deemed competent and thus needs less supervision is dependent on the individual and requires ongoing review by the relevant member of management. The organisation should develop an electronic training database, or a paper-based or electronic training matrix, to control the training requirements for the organisation. The training matrix, or equivalent, should state the current level of training for all staff. It should identify for a specific job role what level of training and competence is required in terms of internal and external training and the current level of training of all staff in that job role. This will assist the organisation in undertaking a training needs analysis to identify those who require further training for the first time or in the form of refresher training. The training matrix should therefore identify those staff currently in the induction phase and receiving training, those individuals who are signed off as competent, but may need refresher training and those who are deemed able to train others. The training matrix should be reviewed at designated intervals. If as a result of review and/or other audit activities inconsistencies in training and competence are identified, appropriate preventive or corrective action needs to be determined and then implemented. The details of the training matrix review and any subsequent preventive or corrective action should be recorded and records retained as a demonstration of due diligence. Formal employee appraisals should also be undertaken. These employee appraisals should include ongoing performance monitoring and should determine if there are any areas where the individual concerned requires further development support and/or training.

Cessation   17.18   Where employment is terminated for any reason, consideration
of Employment   should be given to any possible risks arising through disaffection or simple lack of continuing interest and commitment from the employees concerned, and appropriate precautions taken as necessary.

Food Hygiene   17.19   Many important aspects of hygiene are also dealt with in
and Personal Hygiene   Chapters 10 and 19. It is essential that all personnel are appropriately trained in relevant aspects of food hygiene. **Regulation (EC) No. 852/2004 of 29 April 2004** on the hygiene of foodstuffs, as amended, came into automatic force in all Member States on 1 January 2006, and continued the requirement of earlier legislation that food business operators should ensure:

1.   that food handlers are supervised and instructed and/or trained in food hygiene matters commensurate with their work activity;

2.   that those responsible for the development and maintenance of the procedure referred to in Article 5(1) of this Regulation

or for the operation of relevant guides have received adequate training in the application of the HACCP principles; and

3. compliance with any requirements of national law concerning training programmes for persons working in certain food sectors.

Sections 17.20–17.22 refer particularly to ensuring high standards of personal hygiene of all concerned with the production processes and all persons entering production areas, including quality control personnel, product development personnel, engineers and maintenance personnel, inspectors, senior management, contractors and visitors.

17.20 Instruction should not be regarded as an adequate substitute for training, which provides not only information on what should be done, but an understanding of why it is important. Training should embody the requirements of personal hygiene, the reasons why they are important and the fact that they are legal requirements. Adequate facilities and resources must be provided to enable personnel to comply fully with those requirements.

17.21 In the UK, the manufacturer must comply with the requirements of **EU Regulation (EC) No. 852/2004** and **EU Regulation No. 853/2004**, which are also tabulated in Schedule 2, Specified Community Provisions, in the **Food Hygiene (England) Regulations 2006** and in the equivalent Scotland, Wales and Northern Ireland Regulations (or with any regulations that may at any future time supersede these regulations). **EU Regulation (EC) No. 852/2004** states that every person working in a food-handling area is to maintain a high degree of personal cleanliness and is to wear suitable, clean and, where necessary, protective clothing but does not specify the detail of personal hygiene requirements. However, the following should be regarded as essential:

(a) Every person working in a food-handling area shall maintain a high degree of personal cleanliness and shall wear suitable, clean and, where appropriate, protective clothing. It should be noted that the term 'protective' designates protection of the food from contamination and should not be seen in the context of personnel protection and safety.

(b) No person known or suspected to be suffering from, or to be a carrier of, a disease likely to be transmitted through food or while afflicted, for example with infected wounds, skin infections, sores or with diarrhoea, shall be permitted to work in any food-handling area in any capacity in which there is any likelihood of directly or indirectly contaminating food with pathogenic microorganisms, and a person who suspects that there is any such likelihood on his/her part must report it to the appropriate person within the food business. Further guidance is available from the Food

Standards Agency (FSA) publication *Food handlers: Fitness to work regulatory guidance and best practice advice for food business operators* (2009).[2]

(c) An adequate number of hand-washing stations must be available, suitably located and designated for cleaning and effective drying of hands. All sanitary conveniences within food premises must be provided with adequate natural or mechanical ventilation. An adequate number of flush lavatories must be available and be connected to an effective drainage system. If toilets other than non-flush-type lavatories are used, then a documented risk assessment must be in place to state that the use of alternative designs does not present a contamination risk to food handlers. This risk assessment must be validated and routinely verified to demonstrate that the levels of personal hygiene required in a food manufacturing operation are consistently achieved. Lavatories must not lead directly into rooms in which food is handled, that is, there should be a minimum of two doors with an intervening ventilated space. Signs must be displayed in all toilet areas in a range of languages as required to inform people that they must wash and sanitise their hands before leaving the area. Hand-washing stations with adequate resources must be provided within the toilet areas as well as on entry to production areas. There must be adequate training and ongoing supervision to ensure that protective clothing is removed and suitably protected from contamination prior to entry into toilet areas. If showers are provided, consideration should be given to the potential contamination of shower heads with *Legionella* bacteria. A procedure should be established, implemented and verified to control the potential for *Legionella*.

(d) Hand-washing stations must be provided with hot and cold running water, materials for cleaning and, if required, sanitising of hands, and for hygienic drying. The provisions for washing food and washing equipment must be separate from the hand-washing facility. Taps should not be present. Instead the water flow should be knee, foot or magic eye operated. The turning off of the water must be automatic. Signage should be at the entrance to all production areas requesting that hand washing is undertaken before entry. Signage on best practice for hand-washing should also be in place, including schematics for how to wash, clean and disinfect hands, especially in manufacturing environments where high-risk foods are prepared. Minimum contact times for chemical products on hands should also be identified. Instruction on effective hand-washing techniques should be given during induction and refresher training given as required. Hand-washing procedures should be documented, and personnel should wash hands on return to the

[2] http://www.food.gov.uk/multimedia/pdfs/publication/fitnesstoworkguide09v3.pdf.

production area and before commencing work, after handling waste or undertaking cleaning tasks, after visiting the toilet and after all break periods. Where required, hand sanitiser should be used. The FSA publication *E. coli O157 – Control of cross-contamination: Guidance for food business operators and enforcement authorities* (2014), section 4, recommends that for extra protection against cross-contamination a liquid hand wash that has disinfectant properties conforming to the European standard BS EN 1499:2013 is used. Hand-sanitising gels, if used, should comply with BS EN 1500:2013. Hand-sanitising gels are of limited value on dirty hands as they will not remove dirt and they should not be used as a substitute for effective hand washing first before sanitisation. Consideration should be given to cultural barriers to the use of alcohol-based products and suitable alternatives sought where required. Routine hand swabbing could be used to determine the effectiveness of hand-washing techniques. The temperature of the water used for hand washing should be routinely checked and recorded. The *Campden Guideline G62: Hand hygiene: Guidelines for best practice* (2009)[3] recommends a water temperature of between 35 and 45 °C. Signage should be in place to identify areas designated for hand washing only. The provisions for washing food and also washing equipment must be distinct from hand-washing areas and signed as such. Practice should be routinely verified to ensure that the correct activities are being undertaken at each specific designated sink/washing station and there is no risk of cross-contamination with mixed use of a specific facility. The FSA has produced a series of food hygiene training videos, including information on effective hand washing.[4]

(e) The wearing of wrist watches and jewellery except for plain wedding rings or wedding wrist bands should be prohibited in food production and storage areas. The company should have a defined policy on the type of jewellery allowed to be worn for medical, ethnic or religious reasons and the controls in place to ensure personnel health and safety as well as minimising the risk of product contamination, especially if wedding rings and plain sleeper earrings are worn. Piercings, including rings and studs in exposed parts of the body such as ears, noses, tongues and eyebrows, should not be permitted.

(f) Adequate changing facilities for personnel must be provided where necessary. The changing facilities must be designed for effective cleaning. They must provide adequate space for the number of permanent, agency and temporary staff and be designed to separate outside and inside workwear and provide secure storage for personal belongings, including clothing.

[3] http://www.campden.co.uk/publications/pubDetails.php?pubsID=4480.
[4] https://www.food.gov.uk/business-industry/food-hygiene/training.

There must be provision for designated storage for clean and dirty protective clothing to prevent cross-contamination. Staff should enter the production facility direct from the changing room and not go outside in their protective clothing before entry into the production facility. Visitors and contractors may use the same changing facility or a designated facility; again they must enter the production areas directly from that facility. Lockers must not be used to store food items. A separate provision must be made on-site for storage of food brought onto site. The cleaning of lockers should be addressed in the cleaning schedule and their maintenance should be undertaken as required. Lockers should be so designed that dirt, rubbish and waste material cannot accumulate either on top of or underneath the lockers. Internal protective clothing should not be worn outside and should be removed on leaving a food production area and be suitably stored. Designated separate lockers may need to be provided as necessary to the job role, for both internal workwear (protective clothing) and external workwear, such as high-visibility jackets, as well as personal belongings or clothing that must not be taken into the production and storage areas. In high-care food manufacture, a protocol area is recommended, which requires formal separation of personal clothing, external workwear and internal workwear, including the adoption of separate lockers as well as a barrier in the room to formally segregate high- and low-care areas.

(g) There should be an adequate provision of safety footwear and suitable protective clothing and the laundering thereof. There should be sufficient protective clothing per person to ensure that appropriate changing of clothing can be undertaken. The frequency of changing of protective clothing must be determined by risk assessment. Laundering of protective clothing could be undertaken on-site or by a third-party contract laundry service. The contractor supplying laundry services must be subject to the supplier approval and monitoring procedures (see Chapter 23). Designated internal footwear such as shoes or boots should be segregated from external footwear. In high-care environments, personnel should understand why disinfection of internal footwear is important and follow the designated barrier procedures at all times.

(h) Laundry procedures must provide for the appropriate degree of cleaning of protective clothing. There should be adequate segregation during storage and after laundry of clean and dirty workwear so clean workwear is protected from recontamination. The temperature of the wash/dry cycle should be determined as to whether the items need to be cleaned or in the case of high care, cleaned and disinfected through the temperatures achieved in the washing/drying cycle. Temperatures should be monitored on a routine basis to ensure they meet minimum requirements. The results of monitoring should be recorded. High-risk and low-risk area clothing must be laundered separately. Consideration should be given, based on risk assessment, as to whether home laundry is appropriate. If home laundry is permitted, specific procedures must be put in place and staff should not be allowed to travel to work in their protective clothing or other associated workwear. If company-issued clothing/uniforms are used that are permitted to be worn to and from work, a risk assessment must be undertaken to determine what additional protective clothing is required to protect food and materials from contamination. This is especially important in low-risk businesses that allow home laundry and the wearing of workwear/uniform from home to work, for example in despatch areas. Suitable provision must be made for smoking and eating of food to prevent contamination. Verification activities to determine the effectiveness of the laundry procedures must be implemented and action taken to address non-conformance if required.

(i) Clean protective overclothing, including headgear and, where appropriate, neck covering, must be worn by food handlers and any persons visiting food rooms. Protective clothing must be designed with no external pockets and fastened with press studs rather than buttons as well as falling to below the knee. Specific cleaning and/or disposal procedures must be developed for items of protective clothing, for example plastic disposable clothing, chain mail aprons and gloves, and liner and outer gloves and coats used in chilled despatch and freezer stores. The cleaning of high-visibility clothing and hard hats where they could come into contact with the product must also be determined and procedures implemented. Hoods on underclothing must be contained within the protective clothing at all times as well as any fastenings on underclothing that could pose a food contamination risk if dislodged. Scarves should not be worn in chilled and frozen temperature-controlled areas as there is a personal health and safety risk. Where disposable protective clothing is used, procedures for adequate control should be developed based on risk assessment.

(j) There should be pre-employment medical checks so that no person suffering from, or who is a carrier of, any of the specified kinds of infection is employed as a food handler. Personnel should be actively encouraged to report infections and skin lesions, and supervisory personnel should be encouraged to look out for signs and symptoms of such conditions. Return-to-work procedures following illness or holidays abroad should be developed. Medical screening requirements will vary according to the types of product being manufactured and whether they are high or low risk. Pre-employment medical checks can include stool sampling to verify that individuals are not carriers of harmful pathogens before they commence work or be a simple structured series of questions. Examples of such pre-employment questionnaires are included in the *FSA Food Handlers: Fitness to Work—Regulatory Guidance and Best Practice*

Advice for Food Business Operators 2009, published in May 2009. The quality control manager, or designate, should verify that the person is fit to work as a food handler, both on pre-employment checks and in the case of illness or holidays abroad. The individual should be competent to check the questionnaires and as a minimum hold a current Level 2 food safety certificate.

(k) The carrying of loose items in the production areas, especially mobile phones, should be prohibited. Where deemed necessary, pen control policies, as with any other loose items, for example keys, should be in place to prevent loss or contamination of food.

(l) Eating, drinking and smoking should only be undertaken in a designated area. Protective clothing should be removed before breaks periods. Smoking while wearing protective clothing should be prohibited. If canteen/restaurant or food service facilities are provided either directly or via a catering supplier, there must be a HACCP plan in place that addresses the provision of this food. In the UK the Safer Food Better Business (SFBB) manual may be adopted within the kitchen and food service facility. The FSMS and the associated records for food that is provided for staff should be reviewed on a routine basis by the quality control manager or designate. The personnel hygiene requirements identified in this chapter should apply to the canteen/catering staff as equally as food production staff. If canteen/restaurant facilities are not provided on-site, and staff bring in their own food, they should be provided with adequate designated refrigerated storage facilities that are clean, hygienic and capable of maintaining appropriate temperatures. Fridge air temperature should be checked on a routine basis to ensure compliance. If cooking facilities are provided, including microwaves, they, as well as the fridge, should be identified on a cleaning schedule and a cleaning programme developed, implemented and verified. Particular attention should be given to the cleaning and disinfection of food contact surfaces, which could be vehicles for food contamination.

(m) Perfume, scented deodorant or aftershave should not be worn as this could potentially taint the product. For this reason perfumed disinfectants or hand soaps should also not be used.

(n) All hair should be fully contained to prevent product contamination. This will entail the use of head coverings, hairnets and snoods as appropriate. Consideration should be given to the containment of body hair on arms, torso and so on, and to this end protective clothing should come down to the wrist and fasten to the neck. The *Campden BRI Guideline G48: Guidelines for preventing hair contamination of food* (2006; ISBN 0905942779) provides further information on the protocols that should be in place.

(o) Glove wearing should be controlled to prevent them from being a source of foreign body contamination. A glove issue and control procedure should be in place and awareness training undertaken so that personnel are instructed that the gloves are worn to protect the food from contamination not to keep their hands clean and that a gloved hand is as capable of causing cross-contamination as a bare hand. Latex gloves should not be used due to the risk of allergenic reaction by individuals allergic to latex. If a part of a glove, or indeed the glove itself, is lost, then the quality control manager, or designate, should be notified immediately so that appropriate corrective action can be implemented (see 17.21q for corrective action procedure in the event of the loss of a glove or plaster).

(p) Fingernails should be kept clean, short and unvarnished. False fingernails or nail varnish should not be worn as they could prove a source of foreign body contamination. False eye lashes and/or excessive make-up should not be worn. Consideration should be given, depending on the food products being manufactured, to formal checks being undertaken before work commences to prevent contamination.

(q) All cuts and grazes on exposed skin should be covered with a company-issued detectable coloured metal strip plaster. These must be issued and signed out and checked at the end of production to ensure they are still in place. The batches of plasters, where appropriate, should be regularly tested through a metal detector. Wounds obtained out of working hours (in particular, those to the hands, arms and face) should be reported to the relevant manager at the start of work. Normal plasters must then be removed and replaced with company-issued dressings. First-aid kits must be under formal control and be monitored to ensure that all items stored within the first-aid kit and when they are used can be fully accounted for. The stock in the first-aid kits must be routinely verified to ensure effective control. It is accepted practice to determine at the end of the shift, or day's production as appropriate, that any plasters issued are accounted for. The loss of a plaster, or a piece of or a whole glove must be reported immediately to the relevant manager and the following actions taken:

- production should be stopped;
- a search must be undertaken of the immediate location, general workplace and waste containers for the plaster as appropriate;
- if the plaster cannot be found, then packed product must be re-inspected;
- if the plaster is still not found and there is the potential for the item to be in a consignment that has already been despatched, the quality control manager or designate must be notified immediately and appropriate corrective action implemented.

(r) Staff should inform managers or supervisors if they are required to take personal medicines, especially for long-term conditions such as for allergies, diabetes or asthma. A personal medication procedure should be in place to control all medication brought onto site in order to prevent inadvertent contamination of the product and the procedure must be routinely audited to determine its effectiveness.

(s) Visitors and contractors should undergo medical screening before being allowed entry to production areas. Where pre-entry medical questionnaires are completed by visitors and contractors, they should be formally reviewed and signed off by an appropriate member of staff before the individuals are allowed to enter a production facility. Visitors and contractors should be accompanied by staff, or otherwise controlled, at all times to ensure that personal hygiene requirements and product protection controls are fully complied with and security arrangements are fully met.

(t) Verification activities should be undertaken on a prescribed time interval basis to ensure that personnel are fully complying with all the above requirements. Programmes of refresher training for employees should be undertaken at regular intervals and as required if issues of poor personal hygiene practices are identified.

Personal Hygiene Policy   17.22 It is the responsibility of the quality control manager to ensure that personal hygiene policies and protocols have been developed and that good hygienic practice has been explained to, and is fully understood by, employees, visitors and contractors. Managers should communicate the policies and protocols to all personnel and demonstrate full management commitment to the requirements, advise employees, visitors and contractors of their legal obligations under European Union (EU) and UK food hygiene legislation to report any infectious or potentially infectious conditions, and exclude infectious or potentially infectious food handlers as specifically required by EU legislation. Managers should continuously review whether there are any barriers to good personnel hygiene practice and remove them where necessary. These include ensuring that:

- the design of the premises and positioning of hand-washing stations allows for adequate hand washing and sanitisation if required;
- there are adequate supplies of clean protective clothing and hand-washing sundries at all times;
- staff have sufficient time available to undertake good hygienic practice;
- the personal hygiene training given, including the training aids used, is appropriate to the language skills and literacy of the staff; where appropriate, pictorial training should be used; and
- staff, visitors and contractors have sufficient understanding of the potential effects of poor personal hygiene.

Managers should follow appropriate action, including retraining, with staff who exhibit unhygienic practices.

17.23   The site should develop and implement a sickness reporting procedure, which identifies the actions to be taken should an individual fall ill either before or during his/her period of work. All staff should notify the manufacturing business in the event that they or a person they have been in contact with is suffering from or carrying an illness that could cause a food poisoning or foodborne illness outbreak before they commence work. The *FSA Food Handlers: Fitness to Work – Regulatory Guidance and Best Practice Advice for Food Business Operators 2009*, published in May 2009, identifies best practice in this area. Procedures should be in place should an individual fall ill at work in terms of reporting procedures, and these should form part of induction training. Procedures should also be implemented for managing any spillage of bodily fluid spillages, for example vomiting and bleeding, within the production and storage areas. Bodily spill kits are available that can be used and then disposed of in a way that minimises the risk of contamination. All contaminated products should be disposed of, as appropriate.

Agency Staff   17.24   The employment of agency staff should be in compliance with the legal requirements operating in the country where the manufacturing site is situated. In the UK, employment agencies that operate in food processing must hold a valid Gangmaster and Labour Abuse Authority (GLAA) approval. The onus is on the manufacturer to check at routine intervals that the agencies' approval has not been revoked. In order to manage this effectively, a formal procedure should be in place with responsibilities defined as to who will coordinate and implement the procedure and, if required, appropriate corrective action. The training requirements outlined in this chapter apply as equally to agency staff as they do to employed staff. If training and medical records are held by the agency, they should be routinely verified as appropriate by the quality control manager or designate. If the agency is responsible for the laundering of protective clothing, then they should comply with all necessary controls (see 17.21h for more information).

Ethical Trading Initiative Base Code   17.25   It may be a requirement of retail customers as a prerequisite to supply that the manufacturer complies with the Ethical Trading Initiative (ETI) Base Code, SA 8000 or another third-party standard. It is best practice for manufacturers where this is a requirement to ensure that their suppliers comply with the relevant ethical trading standard(s) as well. It may also be a supply chain requirement for a manufacturer to be a member of the Supplier Ethical Data Exchange (Sedex) as well as having a formal Sedex Members Ethical Trade Audit (SMETA). For further details see Chapter 18.

Food Integrity Management   17.26   Training in food integrity management and to create awareness of the potential for food crime and illicit behaviour is necessary to enable personnel at all levels within the organisation to understand their personal responsibilities and improve their knowledge and skills in this area of GMP. Training to ensure that there is adequate understanding of organisational whistle-blowing procedures should be undertaken and there should be verification activities carried out to measure the degree of understanding at all levels of management. For more details see Chapters 5 to 7.

18   WORKER WELFARE STANDARDS

Principle

There should be policies and procedures that define worker welfare standards for the manufacturing organisation. These policies and procedures should be appropriate to the tasks undertaken, and reflect the requirements of legislation in the country where manufacture takes place and the requirements of supply chain standards where these are higher and a prerequisite to supply. Worker welfare standards where these relate to the working environment, welfare facilities, good housekeeping and physical safety are all key elements of good manufacturing practice (GMP).

Introduction

18.1   The term 'worker welfare' describes the actions taken by a manufacturing organisation to ensure the wellbeing of those who are employed by them and their wider supply base. As described in Chapter 17, the physical health of employees, especially if they suffering from or are a carrier of any potential disease or infection which could via their work activities contaminate other staff or the food product itself, is a key concern that is addressed by GMP (see 17.21 and 17.22). Personal occupational health and safety issues are considered in Chapter 42 of this Guide and again fall within the context of worker welfare. However, worker welfare as an issue extends beyond these two aspects to a far wider scope. Ethical trade standards focus on the protection of workers' health, safety and rights throughout the supply chain and how they are impacted by the contractual interaction of organisations in the supply chain.

Ethical standards can be of two types: business-to-business (B2B), where there is a requirement for a manufacturing organisation to demonstrate that it meets minimum contractual requirements for workers' health, safety and rights, and business-to-consumer (B2C), where the compliance of organisations in the supply chain to ethical standards is communicated to the consumer via labelling of specific products, for example Fairtrade bananas, coffee or chocolate and so on. In terms of GMP it is important for the manufacturer to consider whether there is a contractual requirement to meet legislative requirements alone in terms of worker conditions or whether their customers require them to comply with additional market standards that could be B2B or B2C in orientation.

Business-to-Business Ethical Standards

18.2   It may be a requirement of retail or food service customers that as a prerequisite to supply that the manufacturer demonstrates that they comply with a B2B standard such as the Ethical Trading Initiative (ETI) Base Code,[1] SA 8000:2014 or another second- or

---

[1] https://www.ethicaltrade.org/eti-base-code.

Food & Drink – Good Manufacturing Practice: A Guide to its Responsible Management, Seventh Edition.
The Institute of Food Science & Technology Trust Fund.
© 2018 John Wiley & Sons Ltd. Published 2018 by John Wiley & Sons Ltd.

third-party standard. The ETI Base Code is based on the conventions of the International Labour Organisation (ILO) and is an internationally recognised code of practice. Social Accountability International SA 8000:2014[2] is a standard also based on the conventions of the ILO and the Universal Declaration of Human Rights, and has a series of system standards that organisations are required to meet in order to gain certification.

18.3   It is best practice for manufacturers, where ethical trading is a market requirement, to ensure that their suppliers comply with sustainable sourcing policies as well as the relevant ethical trading standard(s). It may also be a supply chain requirement for a manufacturer to be a member of the Supplier Ethical Data Exchange (Sedex) as well as having a formal Sedex Members Ethical Trade Audit (SMETA).

Business-to-
Consumer Ethical
Standards

18.4   A manufacturer may be required to comply with specific retailer or food service organisation standards for ethical trading where these are then communicated directly to the consumer in terms of labelling on the product. Where claims are made on packaging, the onus is on the manufacturer to demonstrate that the product and the ingredients used to make the product can be traceable to source. The Fairtrade claim, however, may only relate to one ingredient in a composite product and it is important that for those products traceability is assured.

Modern Slavery

18.5   In the United Kingdom (UK), the **Modern Slavery Act 2015** applies to public and private companies if they have a global net turnover of over £36 million and the company carries on all or any part of its business in the UK. Companies who meet this criterion have an obligation to publish a slavery and human trafficking statement every year, six months after the end of the company's financial year. Whilst this may seem from the turnover figure identified to be of interest only to large manufacturing organisations, there is a requirement for large organisations to include information in their review on how the supply chains they interact with meet this requirement too. The Stronger Together Initiative[3] is a multi-stakeholder approach aiming to address modern slavery and human trafficking and provides resources and information for manufacturers who are required to demonstrate that they have considered modern slavery and human trafficking in their own business and their wider supply chain. The resources available include risk screening tools and templates for documents and policies.

Gangmaster and
Labour Abuse
Authority

18.6   In collaboration with other agencies, e.g. HM Revenue and Customs, the Gangmasters and Labour Abuse Authority (GLAA) investigates labour exploitation concerns in the UK and

[2] http://www.sa-intl.org/index.cfm?fuseaction=Page.ViewPage&PageID=1689.
[3] http://stronger2gether.org.

is responsible for regulating the activities of gangmaster, employment agencies and labour providers in the agriculture, horticulture, shellfish gathering, and food and drink processing and packaging sectors. A labour provider must have a GLAA licence to provide labour in these areas of employment.

Agency Staff 18.7 The employment of agency staff should be in compliance with the legal requirements operating in the country where the manufacturing site is situated. In the UK, employment agencies that operate in food processing must hold valid GLAA approval. The onus is on the manufacturer to check at routine intervals that the agencies' approval has not been revoked. In order to manage this effectively, a formal procedure should be in place with responsibilities defined as to who will coordinate and implement the procedure and, if required, appropriate corrective action.

19   PREMISES AND EQUIPMENT

Principle

Buildings should be located, designed, constructed, adapted and maintained to suit the operations carried out in them and to facilitate the protection of materials and products from contamination or deterioration. Equipment should be designed, constructed, adapted, located and maintained to suit the processes and products for which it is used and to facilitate protection of the materials handled from contamination or deterioration. Consideration should be given where appropriate to segregation of personnel and equipment from high- and low-risk food production. Physical separation of high- and low-risk food production and physical segregation of production of foods containing major allergens (see Chapter 8) should also be considered as part of the design process.

General

19.1   Depending on the products being handled, reference should be made to the detailed requirements in respect of premises and equipment in **EU Regulation (EC) No. 852/2004** on the hygiene of foodstuffs, and **EU Regulation (EC) No. 853/2004** laying down specific hygiene rules for food of animal origin (or with any regulations that may at any future time supersede these regulations). This regulation is implemented in England by the **Food Safety and Hygiene (England) Regulations 2013** (as amended) and equivalent legislation in other countries of the UK. An overall site plan should be available that defines the location of the manufacturing unit in terms of buildings and external activities and services, for example chemical storage, waste storage, raw material, packaging and production areas. The neighbouring businesses should also be identified, especially where they could pose a specific food safety risk, e.g. a farm, chemical manufacturer, waste treatment works etc. The site plan(s) should also identify personnel flow around the site, including internal and external access, staff facilities and pedestrian routes to work areas (see 19.18). Areas that may be subject to restricted access should also be highlighted (see 19.2). This is important with regards to food safety risk management and also wider food integrity management (see Chapter 7). High-risk and low-risk product areas for both food safety and food integrity should be identified together with the production and process flow, routes for waste, quarantine areas and the movement of reworked products. The FSA publication *E. coli O157 – Control of cross-contamination: Guidance for food business operators and enforcement authorities* (2014) states that premises should be designed to ensure adequate physical separation; anything else 'will involve a shift towards greater uncertainty regarding the stringency of risk reduction that can be achieved'. Where there is dual use of equipment in manufacturing for both raw and high-risk products or low- and high-risk products, suitable procedural controls must be in place. The procedural controls must be

Food & Drink – Good Manufacturing Practice: A Guide to its Responsible Management, Seventh Edition.
The Institute of Food Science & Technology Trust Fund.
© 2018 John Wiley & Sons Ltd. Published 2018 by John Wiley & Sons Ltd.

validated and then consideration should be 'given to the monitoring and management arrangements required to ensure proper implementation of these procedures'.

19.2   Premises should be sited with due regard for the provision of services needed and to avoid contamination from adjacent activities. In existing premises, effective measures should be taken to avoid such contamination. Any measures introduced should be routinely reviewed to ensure that they remain effective. Examples include the provision of water, ice, compressed air, gas and air supply to the production area. The grounds surrounding the buildings should be maintained to minimise potential harbourage for pests that could be afforded by old pallets, packaging, waste, equipment and machinery or vegetation that is growing up the external walls of the building. Vegetation should be cleared on a regular basis, and there should be a minimum of a 1-metre gap between vegetation and externally stored materials and buildings to prevent pest harbourage. Best practice would be to increase this distance further. Consideration should be given to adequate drainage in the yard and minimising pools of waste water through which vehicles may pass and then enter food manufacturing areas. The potential for flooding should also be considered, especially with regard to sudden storm events. If the drainage system may not cope with such events then appropriate emergency measures should be available in high-risk areas. External traffic areas should be suitably surfaced to prevent contamination or damage (through concussion) of product and packaging. This is especially important when glass product containers are used.

Building integrity should ensure that there are no access points for pests to gain entry or allow ingress of water through seepage or poorly designed guttering. Site security should be reviewed and a risk assessment documented with regard to the potential for malicious tampering or criminal activity with a view to accessing the product and/or process. The need for restricted areas should be considered, including the need for swipe card or keypad entry. The requirement for closed-circuit television (CCTV), infrared heat monitoring, security guards and/or fencing that fully encloses the site should also be considered as well as how staff, visitors and contractors are authorised to enter certain areas. Protocols for lone working should also be considered not only with regard to health and safety, but also with regard to product integrity and the potential for malicious tampering/sabotage. Guard dogs must only be allowed access to external areas and must be under the control of their handlers and not be allowed to run free. If guard dogs are used on the premises, then a risk assessment must be undertaken to determine the level of product risk and associated controls put in place. This must be documented.

19.3   Premises should provide sufficient space to suit the operations to be carried out, allow an efficient work flow and facilitate effective communication and supervision. The need to segregate high- and low-risk operations and those involving major allergens should also be considered. Physical segregation of materials may be required or the premises may need to be designed to ensure that personnel cannot transfer from one area to another during their work activities, for example a raw and cooked meats factory, which requires personnel segregation. Segregation should consider the following: type and hazards associated with raw materials; in-process and finished products, especially the control of allergens; the flow of product, rework, waste, packaging and materials, equipment (including maintenance tools), personnel, airflow, utilities and water; and transfer points between high- and low-risk areas. If physical segregation, including barriers, is used, then consideration should be given to personnel health and safety in the event of an emergency, for example fire. Fire doors should therefore be alarmed or tamper evident so they cannot provide a personnel thoroughfare into and out of food areas. When designing or redesigning the premises, the following may have to be taken into account:

- the organisation must comply with regulations that require the site to be approved or registered in terms of hygiene and premises and processing standards with the local government authority;
- the availability or requirement for services such as electricity, power, sewerage, waste disposal, airflow, refrigeration plant and drainage;
- the requirements for effluent treatment prior to discharge, and provision for waste control, including material segregation for recycling, where deemed necessary;
- the condition of external infrastructure, including concrete and hardstanding and the potential for fugitive release of potential pollutants, and the condition of bunding and containment measures and whether they are suitable for the materials being stored;
- the need to define vehicle and pedestrian routes both internally and externally within the site (see 19.18);
- the availability or requirement for water, ice and/or steam within the process and its suitability for use. The source of the water, for example mains, borehole (well), surface water or recycled water systems, should be considered and its ability to meet potable water standards (see Chapter 20). The requirement for, and control of, cooling water systems should also be considered and the safety procedures that are required, especially with regard to *Legionella*. A water distribution plan should be available that includes all pipework and water-holding tanks as well as the outlining the type of water that is in the location, i.e. potable or recycled. A cleaning schedule should also be in place for the water distribution system that has been validated, is monitored and is routinely verified. The water distribution map can be colour coded to aid differentiation of water systems;

- the raw material/ingredient storage and types of storage required whether frozen, ambient or chilled;
- the retail crate/tray and bulk container (field tray, fruit bin, etc.) storage facilities, especially if this is external. Procedures must be in place to ensure the material is adequately stored to prevent contamination, especially if the product is loose and could come into direct contact with the trays, and that the trays are visually inspected before use;
- the requirement for deboxing/debagging areas for removal of external packaging before items gain entry to the production areas;
- the availability of designated in-process storage and the types of storage required whether frozen, ambient or chilled;
- the suitability of finished product storage prior to despatch and the types of storage required whether frozen, ambient or chilled;
- storage areas should be of sufficient size to enable all operations to be carried out under appropriate conditions;
- the location of inspection and quality control stations within the premises;
- the location of label and packaging printing room (where this task is undertaken off-line);
- the material used for building fabrication and processing equipment. The presence of wood should be minimised and where possible eliminated;
- the production line and equipment layout to seek to maximise efficiency and minimise the risk of product contamination, especially with regard to allergen control and the packing of identity preserved materials such as conventional, assured and/or organic food products;
- the facilities for equipment and premises cleaning and disinfection;
- the cleaning chemical and cleaning equipment storage areas;
- the design of product flow to minimise the risk of contamination and/or cross-contamination;
- the personnel flow in terms of personal health and safety, but also to minimise the risk of contamination and/or cross-contamination;
- the location and siting of hand-washing facilities, especially the location and design of doors after hand washing has been completed in order to access production areas;
- the siting of equipment so that there is adequate access, especially for cleaning and servicing, around it;
- the requirement for charging battery-controlled equipment and the provision of suitable locations, for example charging forklifts away from food environments;
- the restroom and personnel facilities;
- the requirement for protocol areas with barrier control in high-risk food premises; and
- the facilities for maintenance and equipment repairs.

For further information, refer to the *Campden BRI Guidelines for the hygienic design, construction and layout of food processing factories* (2003, Guideline G39; ISBN 0905942574) and the *European Hygienic Engineering and Design Group Hygienic Equipment Design Criteria Guidelines* (2004, or latest version),[1] BS EN ISO 14159:2008 *Safety of machinery. Hygiene requirements for the design of machinery* and BS EN 1672-2:2005 + A1:2009 *Food processing machinery. Basic concepts. Hygiene requirements.*

19.4   All premises, including processing areas, laboratories, stores, passageways and external surroundings, should be maintained in a clean and tidy condition.

19.5   Premises must be constructed and maintained with the object of protecting against the entrance and harbouring of vermin, birds, pests and pets (see also 19.11 and Chapter 22).

19.6   Premises should be maintained in a good state of repair. The condition of buildings should be reviewed regularly and repairs effected where necessary. Special care should be exercised to ensure that building materials and construction, repair or maintenance operations are not allowed to affect adversely product safety, quality or integrity. Buildings should be effectively lit and ventilated to minimise dust and prevent condensation, with air control facilities (including temperature, humidity and filtration) appropriate both to the operations undertaken within them and to the external environment. Where appropriate to the type of food manufacture, positive air pressure systems should be installed (see 19.10).

19.7   There should be sufficient intensity of light to aid the activities being undertaken in the location, for example product inspection and/or cleaning. Areas should be lit to enable personnel to work safely, especially where there is no external light entering the building or personnel are expected to work in the location over a 24-hour period. All light appliances should be suitably protected by either shatterproof plastic diffusers or the use of sleeve covers. Where this is not possible, a fine metal mesh screen must be used. The lights and covers (as appropriate) should be subject to brittle material control procedures (see 19.36).

19.8   Working conditions (e.g. temperature, humidity, noise levels) should be such that there is no adverse effect on the product, either directly or indirectly via the operator, or indeed effects on the staff themselves.

19.9   Fans should be sited to avoid contamination of the product being manufactured and conversely contamination of the local environment. Product contamination could be caused by intake of noxious solids, vapours or gases into the manufacturing unit, or the exhaust of air from the manufacturing unit could contaminate other materials. Due regard should be given to the local

---

[1] https://www.ehedg.org/guidelines/.

environment and the avoidance of nuisance to others, including neighbours to the manufacturing site. Nuisance from the use of fans could include odour, noise and/or dust emissions.

19.10   Air supply and extraction trunking should not introduce contaminants into products. For dry food products, dust extraction equipment may need to be installed. Ventilation and extraction systems must be suitably designed to meet processing requirements in terms of preventing condensation and pest ingress, and managing dust effectively. Systems should be sited to minimise contamination of the product or process. Air that comes into contact with the product should be filtered. The air-handling system must be designed to address:

- the degree of variance in ambient air, for example temperature, humidity, level of dust and particulates;
- the process conditions in terms of temperature, humidity, air containment, air extraction requirements, levels of dust and particulates in the process air; and
- the air filtration cycle in terms of positive pressure, volume of air supply, number of air changes required and air balancing.

Air filtration equipment must be adequately maintained and regularly inspected, and air quality routinely monitored. A risk assessment should be undertaken to determine the need for positive air pressure between high- and low-risk areas, the filter size/grade required to control airborne contamination and the potential for contamination, for example the product residency time in that area. If air socks are used, they must be inspected, cleaned and maintained at a designated frequency based on risk assessment. For further details, consult the air filters for general ventilation **BS EN ISO 16890:2016** series of standards and the *Campden BRI Guidelines on air quality standards for the food industry* (2nd edition) (2005, Guideline G12; ISBN 0905942736).

19.11   Floor and wall surfaces, and all surfaces (including surfaces of equipment) in contact with food must be maintained in a sound condition, and be easy to clean and, where necessary, disinfect. Walls should be sound and finished with a smooth impervious and easily cleaned surface. Where walls are in areas of high traffic movement and there is the potential for damage, crash barriers should be installed. Floors in manufacturing areas should be made of impervious materials, laid to an even surface and free from cracks and open joints. They should be of adequate construction and material for the wear and tear and conditions of manufacture encountered. Floors should be designed so that liquid does not collect in certain areas, and the fall of the floor should be such that any water or waste product travels easily to a suitable drain (see 19.13). For further details, consult the *Campden BRI Guidelines for the design and construction of floors for food production areas* (2nd edition) (2002, Guideline G40; ISBN 0905942566) and *Guidelines for the design and construction of walls, ceilings and services for food production areas* (2nd edition) (2003, Guideline 41; ISBN 0905942590).

Ceilings should be so constructed and finished that they can be maintained in a clean condition. Voids above false ceilings should be regularly cleaned and inspected so that they do not provide harbourage to pests. It is essential to ensure effective seals to walls and floor. The coving of junctions between walls, floors and ceilings in critical areas is recommended. Processing areas should be designed without windows. However, where windows are present in production and storage areas, they should be of toughened glass or plastic, protected against breakage, adequately screened to prevent pest entry and secured, and if ledges are present they must slope away from the glazing. Plastic film can be adhered to internal window surfaces as an extra control in the event of damage or breakage.

Doors, dock levellers and door frames should be of impervious, non-corrodible material, smooth, crevice-free and easily cleanable, and suitably protected to prevent ingress of pests when opened (see Chapter 22). Doors must be monitored to ensure that they remain close-fitting and provide adequate pest-proofing. External doors, especially pedestrian doors, should be designed to be automatic closing doors and, if deemed appropriate, air curtains should be installed. Consideration should be given to whether the design of doors is suitable for food preparation areas, especially where high-risk products are produced. The design of personnel doors should minimise handles that can be a source of cross-contamination between personnel, have kick plates to prevent damage when opening, be seamless where possible to prevent bacterial harbourage and be water resistant in areas of water use. Consideration should be given to whether sliding doors or roller doors are used. Roller doors are in contact with the floor and on lifting, depending on the food environment, can drip material onto product or equipment as it passes underneath. If roller doors are used, a risk assessment should be undertaken to determine the degree of risk of product contamination with such doors being in place. If sliding doors are used, it should be determined if there is any health and safety risk to personnel if they are working in close proximity to the doors when they are opening or closing. Automatic doors should be open for an appropriate time that allows access but prevents ingress of pests and other potential product contaminants. Where strip curtains or similar designs are used, they must be maintained so that they are effective against pest ingress, be cleaned at a prescribed frequency and otherwise controlled to prevent product contamination. Training of staff should include effective door management in order to prevent product contamination, ingress of pests and loss of temperature control.

Materials used for construction of ceilings, doors and doorways should be chosen to avoid tainting or otherwise contaminating food materials. More generally, for both food contact and non-food contact surfaces such as walls, doors and ceilings, painted surfaces are not ideal because flaking may occur, which could pose a potential food contamination risk. Paint, where it cannot be avoided, should meet designated standards. Specialist paint is available for the food industry and advice should be sought. Relevant standards include BS 7557:1992 *Specification for limits of metal release from painted surfaces of articles, liable to come into contact with food.*

19.12 Pipework, suitably protected light fittings, ventilation points and other services in manufacturing areas should be of a material suitable for purpose and appropriate to the area where the services are located, and be sited to avoid creating recesses that are difficult to clean. The material should allow for cleaning and, where required, disinfection (see Chapter 21). Services should preferably run outside the processing areas. They should be sealed into any walls and partitions through which they pass to prevent pest ingress. Intake points where material is received from external deliveries through pipework, or other means, to internal tanks must be suitably designed and operated to be easy to clean, and where necessary disinfect, and prevent ingress or harbourage for pests. End caps should be fitted to all external points, or they should be otherwise enclosed, to prevent contamination, theft and/or malicious tampering (see Chapter 5). When not in use, end caps should be locked off to prevent entry or ingress. Fabrication joints must be sealed and not pose a product risk. The type of welding used should be appropriate for the product being manufactured. Pipework should be checked on a regular basis to ensure full integrity and that any potential for harbourage of harmful bacteria is minimised. Pipework should also be assessed for the potential build-up of biofilms (see 21.13). If biofilms are identified, then appropriate action should be taken.

19.13 Drains should be of adequate size for the manufacturing operations undertaken and should have trapped gullies and proper ventilation. Any open channels should be shallow to facilitate effective cleaning. Machinery and sinks should be sited to ensure that process materials and other waste, including water, is discharged directly to a drain. Drains must flow from high- to low-risk areas of the manufacturing site. A system must be in place to prevent back flow of liquid and air from drains. The siting of drains from laboratories should be considered to determine the potential for product contamination, and where required appropriate action taken to eliminate risk. Ideally, this review should be undertaken at the design stage of the manufacturing unit. A drainage plan should be available that maps both internal and external drainage on the site. This should be updated as changes occur and reissued. Based on current knowledge and the need for ongoing risk assessment in the future, drainage surveys, including the potential use of cameras, is needed to determine if the drainage system links to the treatment process prior to discharge and to confirm the flow of foul (dirty) and surface water drainage, and also to determine the current and future condition of the infrastructure, and the potential for seepage from drains and underground storage tanks into soil and groundwater. This is especially important if groundwater is extracted from below the manufacturing site and is then used as a potable water source for food production and staff facilities.

19.14 An adequate number of flush lavatories must be available and connected to an effective drainage system. Lavatories must not lead directly into rooms in which food is handled or stored. The FSA publication *E. coli O157 – Control of cross-contamination: Guidance for food business operators and enforcement authorities* (2014), Section 4, recommends that for extra protection against cross-contamination a liquid hand wash that has disinfectant properties conforming to the European standard BS EN 1499:2013 is used. Hand-sanitising gels can provide an additional level of protection when used after effective handwashing. Hand-sanitising gels should conform to the BS EN 1500:2013 standard.

19.15 Protection from the weather should be provided for receiving and despatch areas, and for materials or products in transit. External material or product storage is not recommended, but where external storage is necessary, procedures must be in place to minimise deterioration and the risk of contamination, particularly pest infestation.

19.16 Where raw materials or packaging materials arrive in external packaging, a separate deboxing/debagging area should be provided where the packaging may be removed before the materials enter the production area.

19.17 Waste material should not be allowed to accumulate. It should be collected in suitably constructed and identified receptacles, designated for the purpose, for removal to collection points outside the buildings. Waste material should be disposed of at regular and frequent intervals. Disposal of printed packaging materials or raw materials and rejected products should be carefully controlled (see Chapter 24).

19.18 Design of premises should provide separate routes of entry and movement for vehicles and personnel. Designated walkways should be marked in internal and external areas. Where walkways or steps cross over production lines, storage areas or workstations they must be fitted with backplates and/or be suitably enclosed to prevent contamination of the product, packaging or people beneath. The material utilised to make the stairs must be suitably controlled to prevent contamination. Ideally, wood would not be used in such areas. If wood has been used, an appropriate risk assessment must have taken place to determine the degree of contamination risk. The use of ladders or scissor lifts in production and processing areas should also be considered and appropriate action taken to minimise the potential for contamination of both the physical manufacturing area and the food being manufactured itself. When designing premises, it must be considered that manufacturing areas should not be used as a general right of way for personnel or materials, or for storage. Sufficient storage space must be available so that material storage at no time prevents or limit access to fire exits. This would not exclude materials temporarily standing for a brief time in transit from storage to a production area, but its temporary standing should not preclude safe exit from the building if required. The space required for temporary storage must also be considered in premises design and appropriate space provided.

19.19 All operations should be carried out on the manufacturing site in such a way that the risk of contamination of one product or material by another is minimised.

19.20 There should be documented cleaning procedures and cleaning schedules for external, manufacturing and storage areas (see Chapter 21).

19.21 Vacuum or wet cleaning methods are to be preferred. Compressed air, hoses, pressure cleaners, brooms and brushes should be used with care so as not to incur the risk of product contamination. Where compressed air is used, the potential for air contamination, e.g. with oil, should be considered and appropriate action taken.

19.22 Surfaces in contact with food should be inert to the food under the conditions of use and should not yield substances that might migrate or be absorbed into the food. A certificate or other written declaration of compliance should be held for all food contact surfaces to demonstrate that the materials are 'food grade' and comply with **Regulation (EC) No. 1935/2004 of the European Parliament and of the Council of 27 October 2004 on materials and articles intended to come into contact with food, Regulation (EC) No. 10/2011 on plastic materials and articles intended to come into contact with food, as amended, Regulation (EC) No. 450/2009 on active and intelligent materials and articles intended to come into contact with food**, and **Regulation (EC) No. 2023/2006 on good manufacturing practice for materials and articles intended to come into contact with foods, as then amended**. These regulations lay down the basic rules necessary for testing migration of the constituents of plastic materials and articles intended to come into contact with foodstuffs and all material testing must comply with the requirements laid down in this directive or as subsequently superseded. The list of groups of materials and articles involved (see **EC No. 1935/2004 Annex I**) include not only food grade material used in equipment design, but the following 17 materials and processing aids:

1. active and intelligent materials and articles
2. adhesives
3. ceramics
4. cork
5. rubbers
6. glass
7. ion-exchange resins
8. metals and alloys
9. paper and board
10. plastics
11. printing inks
12. regenerated cellulose
13. silicones
14. textiles
15. varnishes and coatings
16. waxes
17. wood.

Compliance with this legislative requirement should be a prerequisite that is built into the process design phase of a new process line or piece of equipment. (For more information on specific migration levels associated with food contact materials see Chapter 24). Surfaces should be durable and able to sustain the activities undertaken without cracking or deterioration, for example inspection tables, conveyor belts and elevators. Where food contact surfaces are liable to wear, they should be of a contrasting colour to the food and raw materials used in manufacturing and should be inspected regularly for evidence of damage.

19.23   Any surfaces in contact with food should be easily cleanable and disinfected, smooth, washable, corrosion resistant and made of non-toxic materials. Surfaces must also be non-porous so that particles are not caught in microscopic surface crevices and become difficult to dislodge. Welds should be continuous, and all welds and joints should be smooth and impervious (see 19.12). Routine swabbing of food contact surfaces should be undertaken to determine the efficiency of cleaning and disinfection activities.

19.24   Surfaces in contact with food should be readily accessible for manual cleaning and disinfection or if not readily accessible, then easily dismantled for manual cleaning and disinfection. If clean-in-place (CIP) techniques are used, it should be demonstrated by the food business operator that the results achieved in terms of cleaning and disinfection without disassembly are the equivalent of those that would be obtained with full disassembly and manual cleaning, and can ensure safe food. The type of food contact surface should also be considered when designing the cleaning and disinfection programme, i.e. flat surfaces without joints or seals between sections will be easier to clean than pipework or moving elements.

19.25 Interior surfaces in contact with food should be so arranged that the equipment is self-emptying or self-draining.

Equipment

19.26 All equipment should be purchased in accordance with specified requirements. It should be suitably tested and commissioned before the line is approved for full-scale production. This could include small production runs with a microbiological sampling plan and shelf-life testing to validate key processes (see 3.12). Equipment should be cleaned and serviced before use and any faults rectified. During commissioning, equipment should be tested to ensure that it meets the purchasing specification, is fit for its intended purpose and can be effectively cleaned, and where required disinfected, and maintained.

19.27 Equipment should be so arranged as to protect the contents from external contamination and should not endanger any materials or product through contamination from leaking glands, condensate, lubricant drips and the like, or through inappropriate modifications or adaptations.

19.28 Exterior surfaces of equipment not in contact with food should be so arranged to prevent harbouring of soils, microorganisms or pests in and on the equipment, floors, walls and supports.

19.29 Equipment should be designed and constructed to allow efficient cleaning and maintenance. There should be detailed written instructions for cleaning and disinfection. Specified materials, methods, safety precautions and suitable facilities should be provided (see Chapter 21).

19.30 Plant and equipment should be cleaned and serviced immediately after use. Any faults should be recorded. Missing parts, such as nuts, bolts, springs and clips, should be reported immediately to the quality control manager or designate.

Maintenance

19.31 The structure of the building, including walls, floors, ceilings and doors, should be maintained in a sound condition (see 19.11). Plant and equipment should be checked for cleanliness and integrity before every use and to this end should be designed with sound, secure, quick-release systems for inspection and disassembly. Appropriate precautions for ventilating fumes from power-driven equipment, heaters and so on should be taken with consideration for both product safety and personnel health and safety. Temporary repairs should be controlled to ensure they do not impact on the safety or legality of the product. Temporary engineering measures such as by-pass pipework should be permanently repaired as soon as feasible, subject to processing constraints.

19.32 A documented maintenance procedure should be developed that addresses both preventive and responsive maintenance. The procedure should be based on risk assessment and ensure that all servicing and breakdown work undertaken on equipment is carried out, that food safety, legality and quality requirements are achieved and that any potential risk to the product is minimised. Preventive maintenance should be considered for all equipment and components that contribute to product safety. The frequency of servicing should be determined by the maintenance manager using factors such as manufacturer's recommendations, equipment and plant history in terms of failure or breakdowns, proposed hours of use and production requirements. This should be documented on a planned maintenance schedule. The planned maintenance schedule needs to be reviewed on a prescribed frequency and amended as required. It should include all equipment and plant in the manufacturing unit and storage areas that ensures that the environment and the food products are manufactured to comply with legislative requirements. The fabric, plant, equipment and machinery should be serviced/maintained according to the maintenance schedule and any actions required documented in a machine log or other such internal document. This will then provide a service history for the piece of equipment, plant or area. The planned maintenance schedule may need to be updated in the event that hygiene audits or other internal audits identify problems with poor maintenance, for example cracking floor joints, peeling or flaking paint, damage or wear. Maintenance and machine servicing contractors must be made aware of the organisation's maintenance procedure before they commence work, and particular attention must also be paid to personnel hygiene, premises hygiene and food contamination procedures that are in force to ensure the contractors' compliance at all times. Waste control procedures must also be formally agreed with contractors to ensure adequate control and compliance with relevant legislation and to minimise the risk of product contamination.

19.33 Responsive maintenance may be required during production following identification of a machine fault. Minor adjustments could be carried out without interruption to the production line where the appropriate personnel have deemed that it does not present a risk to the safety and integrity of the product. Major repairs or adjustments would require the production line/area to be closed down and if necessary screened off. Any parts that are removed should be logged and accounted for by maintenance personnel prior to the line starting up again or the area going back into production. A cleaning procedure should be defined for the activities to be undertaken before a machine/production line is deemed suitable to go back into production. A hygiene clearance production record where the equipment is formally "signed off" back to production for use should be maintained detailing, as applicable, the equipment, brief details of the task undertaken and any parts that have been removed or replaced. The record should be signed off by the maintenance personnel concerned to confirm their work has been completed and by production personnel to authorise, on that basis, the recommencement of production. All maintenance and servicing activities, including the use of contractors, must be undertaken by competent individuals who can demonstrate by training or other means that they are competent to undertake the task required and are aware of and comply with internal organisational procedures.

19.34   Consideration should be given to the potential for cross-contamination between low- and high-risk areas by maintenance personnel, their tools and equipment. Consideration should also be given to managing the contents of tool boxes and similar storage equipment belonging to maintenance staff to ensure that all items taken into the production area are removed and/or accounted for before production can commence/recommence. Tools must present no contamination risk to the product and must be clean and, where required, disinfected and well maintained. They should be checked after use for any signs of damage and deterioration and appropriate action taken in the event that there is concern of a risk of product contamination. There should be information available that demonstrates that all materials, oils and lubricants utilised during maintenance are food grade and do not present a risk to product safety.

19.35   All engineering workstations/workshops must be controlled so that they do not present a risk to product safety or legality. They must be maintained in a suitable hygienic status, and controls should be put in place such as the use of swarf mats and other waste control measures to reduce risk to product safety.

Brittle Material Procedures   19.36   Wherever possible, the use of glass should be avoided and suitable alternatives sought. There are three different sources of glass: glass packaging, fixtures and fittings such as glass windows, light bulbs and dials, and imported glass such as spectacles, watches and drink bottles. Windows should be secured and laminated to contain any glass in the event of a breakage. Light bulbs and fluorescent tubes should be completely covered to contain any pieces of glass in the event of shattering, or be made of shatter-resistant material. A register should be maintained, and updated as necessary, of all glass, ceramic and hard plastic items in the production area and more widely on the manufacturing site that are deemed following formal risk assessment to present a product contamination risk. A risk assessment should be undertaken to determine the frequency of inspection of all items whether each shift, daily, weekly or other prescribed interval. Items should be regularly inspected and the results of the inspection recorded with any required corrective action. This corrective action must be followed up to ensure completion and the removal of a potential product contamination risk. The use of glass and ceramic storage containers should be prohibited in all production areas. If glass or ceramic packaging is used for the final product, then breakage procedures should be in place for the packaging itself as well as other items of risk, for example these procedures should encompass as a minimum delivery, storage, transfer, dispatch and visual inspection on-line and filler breakage procedures. A written procedure must be in place for the replacement of brittle items detailing the measures that are in place to prevent breakage and to minimise product contamination.

Consideration should be given to the use of equipment/items that are made with brittle materials being present in offices, quality control work stations or rest rooms that open directly into storage and production areas. A risk assessment should be undertaken and recorded that outlines the potential for contamination and the controls in place to minimise the risk of occurrence and the risk of contamination.

19.37   Any breakage of glass lenses in spectacles should be treated as a glass breakage incident and must be controlled by relevant procedures. Procedures should also be in place in the event that a contact lens is lost by person(s) on site where this poses a risk to product.

19.38   All personnel must report immediately to their line supervisor or management any broken or damaged glass/hard plastic or ceramic items. This applies to any location within the production unit or site. Any incident of broken glass or hard plastic components on processing equipment or other glass, ceramic or hard plastic breakage that could in any way have affected the product should result in production being stopped and the product immediately held. Any product/part-processed product or ingredients that could be contaminated must be rejected and a thorough search of the area made before any more product is packed/processed. The area and all equipment within a designated predetermined radius of the breakage incident must be isolated immediately and thoroughly searched for fragments. The size of the area should be determined by risk assessment.

19.39   Depending on the type of material of concern, all personnel in the area at the time of the incident should have their clothing and the soles of their footwear checked for glass, ceramic or hard plastic fragments before leaving the area. All remaining fragments should be removed immediately using dedicated equipment for removing brittle material and disposed of carefully. Colour-coded equipment is often used, for example red bins and red brushes and equipment, which is designated 'GLASS ONLY'. Production equipment that may have been affected must be dismantled for in-depth inspection and cleaning. Broken or cracked windows should be removed from the outside, with heavy-duty polythene sheeting taped to the inside of the production area to prevent glass spillages. The glass must be replaced with a suitable alternative. Where practical, fragments should be pieced back together to try and account for all the pieces of glass, ceramic or hard plastic. A sample fragment should be retained for reference in the event of a subsequent customer complaint and for further analysis if necessary.

19.40   The area must be closely inspected again after cleaning and when the area has been declared free of glass, ceramic or hard plastic formal documentation should be completed and signed off by the appropriate manager to formally clear the area for recommencement of production.

19.41   For glass/hard plastic or ceramic breakages in areas remote from the production and storage area, for example offices and restrooms, a thorough inspection of the area should still be made by appropriate personnel in order to assess the potential risk to product. The breakage procedure must be followed as stated above.

19.42   Where other brittle items present a potential contamination risk to the manufacturing process, they should be addressed by a similar procedure.

Knife/Scissors/Blade   19.43   Knives, scissors and blades are generally used for the following
Procedures   activities:

- opening outer packaging;
- cutting through shrink wrap and cardboard;
- cutting packaging fastenings; and
- food preparation.

Food contact cutting equipment should be identified, made of food-grade material and kept separate from equipment used for other activities, for example designated colours for specific activities or departments, such as goods inwards, production or quality control. Knives (and all other sharp equipment) should be company issued and only be allowed in defined areas. Special protected blades are ideal, which are designed for safety and do not have an exposed blade. Knives should be controlled by suitable documented procedures, including the signing out of equipment to an individual and the signing back in again at the end of the production shift on the equipment's return. Where possible, knives should be chained or otherwise securely fastened in a location to prevent their unintentional loss. Knives should be uniquely identified so that they can be assigned to individuals. On return, they should be checked to ensure there are no signs of damage and the blade is intact. Where appropriate, knives should be sterilised to minimise cross-contamination. In the event that a knife is 'lost' or unaccounted for, the management of incidents procedure should be implemented (see Chapter 27).

Consideration should be given to monitoring the condition of knives that form an integral part of machinery. They should be inspected on a regular basis for integrity, and this inspection should be recorded. Consideration should also be given to implementing metal detection as a measure to ensure the potential for metal contamination is adequately controlled.

Where knives are used in the preparation of food items during manufacture, for example vegetable and meat trimming, due consideration should be given to implementing an effective disinfection programme. In order for such treatments to be effective, a cleaning procedure must be in place to remove soil and organic matter, otherwise disinfection by hot water above 82 °C or the use of chemical disinfectants will not be effective. Routinely the effectiveness of the cleaning and disinfection process should be verified by swabbing the knives. It should be noted that spores produced by pathogenic organisms may survive disinfection and present a food safety risk. This risk is both process and product specific.

Risk Assessment   19.44   The risk of contamination from premises and equipment has been explored in depth in this chapter. The need for appropriate risk assessment and mitigation processes has also been outlined. This risk assessment approach must be inclusive and utilise experience across the management team and staff within the manufacturing operation. Risk assessment at a formal level is important, but risk assessment by individuals as part of routine work tasks on a day-to-day basis is also essential as formal processes can never be exhaustive in terms of identifying the hazards that can occur and the remedial action that then needs to be taken to prevent contamination should the event arise. As such training of all staff needs to address their need to be vigilant with regard to the potential contamination of food products from the premises itself and equipment.

20   WATER SUPPLY

Principle

An adequate supply of potable water should be available in the manufacturing unit. Potable simply means 'safe to drink'. Provision should be made for appropriate facilities for temperature control, storage and distribution of potable water. Dead legs should be eliminated in water systems, and if cooling towers and/or showering facilities are used, consideration should be given to monitoring for Legionella.

General

20.1   **EU Regulation (EC) No. 852/2004** (as amended) on the hygiene of foodstuffs requires an adequate supply of potable water. Potable water must be used to ensure foodstuffs are not contaminated by the source of water used. Ice used in the manufacturing process must be made from potable water to ensure foodstuffs are not contaminated. Ice must be made, handled and stored under suitable conditions that protect it from all contamination. Steam used directly in contact with food must not contain any substance that presents a hazard to health or is likely to contaminate the product. **EU Regulation (EC) No. 852/2004** (as amended) on the hygiene of foodstuffs states that water unfit for drinking that is used for the generation of steam, refrigeration, fire control and other similar purposes not relating to food must be conducted in separate systems, readily identifiable and have no connection with, nor any possibility of reflux into, the potable water system. A water distribution map should be available and appropriate controls should be in place (see 19.3). A cleaning schedule should also be in place for the water distribution system that has been validated, is monitored and is routinely verified. The water distribution map can be colour coded to aid differentiation of water systems.

20.2   Water supplies must be potable and be drawn from mains water or otherwise treated to meet potability standards. Routine microbiological sampling should be undertaken to ensure that the water meets these standards and appropriate action undertaken if results indicate potability standards have not been met. Potable water should be as a minimum standard as specified in the latest edition of the World Health Organisation (WHO) *Guidelines for Drinking Water* (currently 4th edition, 2011). Potable water used for manufacturing purposes, including that used in making up products or likely to come into direct contact with the product, must be of potable quality and free from:

(a) any substances in quantities likely to cause harm to health;
(b) any substance at levels capable of causing accelerated internal corrosion of metallic containers and closures causing taints; and
(c) harmful microorganisms.

Food & Drink – Good Manufacturing Practice: A Guide to its Responsible Management, Seventh Edition.
The Institute of Food Science & Technology Trust Fund.
© 2018 John Wiley & Sons Ltd. Published 2018 by John Wiley & Sons Ltd.

The **Water Supply (Water Quality) Regulations 2016** state that water should not contain any microorganism or parasite or any chemical (other than a parameter listed in Schedule 1) at a concentration or value which constitutes a potential danger to human health. To maintain these standards, the water should, if necessary, be chlorinated or otherwise adequately treated. If water is treated, there should be a procedure in place to ensure that the treatment of water is adequately controlled to prevent food contamination. Monitoring must be adequate to control any trends towards a critical limit, especially if chlorine control is deemed within the hazard analysis critical control point (HACCP) plan as a critical control point (CCP) as described in Chapter 3. It is important in this context to differentiate between determining critical limits for both free and total chlorine. All critical limits should be validated before the HACCP plan is implemented. Corrective action procedures must be in place if target levels or indeed critical limits are breached, including consideration of appropriate disposition of product. Public water providers can supply certificates of potability, and these should be retained by the manufacturer and reviewed for any trends that could impact on water suitability. It is important to ensure that the certificates supplied by public water providers are not generic across a wide water distribution network, but actually relate specifically to the local water distribution system from which the manufacturer receives water.

Although the water may be delivered to the manufacturing site at a potable standard, this does not mean that at the point of use it is still at the self-same standard. Consideration should be given to the number of individual usage points and therefore the number of sampling points and frequency of sampling required to verify consistent standards. A risk assessment should be undertaken to determine the level of risk at the point of water use with regard to food safety and quality. The nature of the process is also important and whether the product comes into direct contact with water, or water is a product ingredient. A further consideration is if the product comes into direct contact with the equipment, conveyors and so on as this will increase the risk of contamination. Further factors to be reviewed include, but are not limited to, the original source of the water, procedures for storage of water on-site, materials used in water storage tanks and transfer pipework and whether they predispose the system to the development of biofilms or contamination of the water, the water treatment process that is undertaken prior to use, especially if the treatment itself could cause a potential food safety hazard; and the management of waste water.

20.3   Water that is recycled for reuse should be treated so that it does not pose a potential contamination risk to the premises or product. The treatment process should be routinely monitored to determine its effectiveness. If there are both potable and recycled water systems in the manufacturing unit, they should be suitably identified to avoid confusion so that operatives are aware at all times whether water at an outlet point is potable/non-potable. Water systems may be differentiated using colour coding on taps, coloured hoses or designated water site mapping as previously described (see 20.1).

Boil Water Advice   20.4   In the event of microbiological contamination of mains water supplies, water utilities in the area(s) concerned would, in most instances, issue advice to boil water before use. The management team, with advice from the quality control manager, would have to determine whether it was possible to continue manufacture of food products under such circumstances. Further details can be found in the following publications:

- Institute of Food Science & Technology (IFST) *Advisory Statement on Contamination of Water Supplies: Boil Water Advice*, January 2004;
- *Water Quality for the Food Industry: Management and Microbiological Issues* (2000), Guideline No. G27, Campden BRI, ISBN: 0905942310;
- *Guidelines on the reuse of potable water for food processing operations* (2012), Guideline No. 70, ISBN 9780907503736;
- The **EU Council Directive 98/83/EC** on the quality of drinking water for human consumption as amended by regulations **1882/2003/EC**, **596/2009/EC** and **2015/1787/EC**; and
- http://ec.europa.eu/environment/water/water-drink/index_en.html.

Water Sampling   20.5   Water should be routinely sampled both at the point of entry to
Procedure   the site and at various points around the premises. This will give a good indication of the microbiological loading on entry to the site and whether the water system itself is increasing the bacterial loading of the water. If this is found to be the case, then remedial action will need to be taken to reduce the bacterial loading. The quality control manager should develop a water sampling proto-col based on risk assessment. The risk assessment should consider the frequency of sampling required for both microbiological and chemical analysis of water to verify that it meets minimum potable standards. The risk assessment should also consider the proportion of the finished product that constitutes water added during the manufacturing process such as in the case of diluting fruit juice concentrate in order to pack single strength juice. The inclusion of water where water forms a constituent of the prod-uct, for instance carbonated drinks, or meat preparations where the meat contains added water, means that the frequency of water sampling as defined by risk assessment may well be much higher than when water is utilised only for cleaning and disinfec-tion activities. Where the water source is from a borehole (well) or surface water, levels of nitrate, arsenic and pesticides need to be determined so that the source can be initially validated and then routinely verified. The use of recycled water needs to be considered in the risk assessment and appropriate frequency of

sampling determined to ensure that water treatment has been adequate. Ice should also be sampled and tested, both ice manufactured on-site and ice purchased and delivered to the site.

20.6 Samples may be analysed by an in-house or third-party laboratory. If a third-party laboratory is used, it should be formally approved (see Chapters 23 and 38). Sterile minimum 500-ml sample bottles should be used, and water should be run for a period of time before the sample is taken. Any contamination of the sample bottle and the lid during sampling should be avoided. Ideally samples should be under adequate temperature control during storage and transit and be with the laboratory within 4–8 hours of being taken. The samples should be kept at a temperature of between 2 and 10°C while in transit and storage, and ideally in the dark. This may require the use of a hygienic cool box and/or ice packs.

20.7 If the water is chlorinated, sodium thiosulphate needs to be added to the sample bottle to neutralise any residual free chlorine, so it is important to agree with the laboratory the type of sample bottles required.

20.8 When the results are received from the laboratory, they should be reviewed to check for compliance with the specification. If the results are out of specification, this must be referred to the quality control manager and appropriate corrective action taken. This could include a review of product status, the potential for product contamination and the need for a product recall.

21 CLEANING AND SANITATION

Principle

Cleaning and sanitisation programmes are a key prerequisite to ensure the adoption of good manufacturing practice (GMP). Cleaning will not be effective unless it is fully supported by management. The role of management is to define the hygiene standards required and to communicate these effectively to staff, usually by means of a comprehensive cleaning schedule and associated task procedures. Management must demonstrate commitment by providing the appropriate means, that is, the equipment, the chemicals and the training for staff.

General

21.1 **EU Regulation (EC) No. 852/2004** (as amended) on the hygiene of foodstuffs states that:

1. Food premises are to be kept clean and maintained in good repair and condition.

2. The layout, design, construction, siting and size of food premises are to:
    (a) permit adequate maintenance, cleaning and/or disinfection, avoid or minimise air-borne contamination, and provide adequate working space to allow for the hygienic performance of all operations;
    (b) be such as to protect against the accumulation of dirt, contact with toxic materials, the shedding of particles into food and the formation of condensation or undesirable mould on surfaces;
    (c) permit good food hygiene practices, including protection against contamination and, in particular, pest control.

21.2 Cleaning procedures must not only be consistent with food hygiene legislation but also with the requirements of environmental and health and safety legislation (see Chapters 41 and 42), and cleaning and disinfection must be undertaken using procedures designed to minimise the risk of product contamination.

Cleaning Schedules/ Procedures

21.3 Cleaning schedules coordinate cleaning activities and are a means of information transfer between management and staff. They should be established for external areas, the premises itself and equipment, plant and services. They should also be established for transport vehicles in the distribution supply chain. Individuals who have the appropriate level of competence and knowledge to develop an effective cleaning and disinfection regime should design cleaning schedules and associated procedures. The site hygiene plan and associated cleaning schedules and cleaning procedures should be based on a risk assessment approach (see 21.7). Documentation such as cleaning schedules and associated task procedures must be clear and unambiguous.

Food & Drink – Good Manufacturing Practice: A Guide to its Responsible Management, Seventh Edition.
The Institute of Food Science & Technology Trust Fund.
© 2018 John Wiley & Sons Ltd. Published 2018 by John Wiley & Sons Ltd.

Consideration must be given to the language skills, numeracy and literacy of staff. Cleaning schedules and associated task procedures must state as appropriate:

- what area or equipment is to be cleaned;
- who is to undertake the task (i.e. responsibility);
- when it is to be cleaned (i.e. how often);
- how it is to be cleaned (i.e. work instruction or task procedure, including how the equipment should be dismantled and reassembled);
- the time duration of the cleaning task, especially the required contact time for any chemicals used. Contact time is the time that the cleaning chemical has to be in contact with the surface in order to be effective;
- the materials to be used, that is, the cleaning chemicals and cleaning equipment (see 21.6);
- dilution rates and whether the chemicals are pre-diluted or need to be diluted by staff to the correct rate at the time of the cleaning operation;
- the requirements for rinsing between chemicals and if required the protocol for the pH testing of rinse water to ensure effective rinsing of equipment has taken place;
- the cleaning standard expected in terms of either visual standard or if microbiological or adenosine triphosphate (ATP) sampling is undertaken the standards to be reached to approve the cleaning task as having been effective;
- the operator personal safety aspects that need to be addressed to ensure that the personal protective equipment (PPE)/clothing that may need to be worn when handling concentrated chemicals and/or when completing cleaning tasks is defined;
- the health and safety of staff and the environmental concerns in the event of a spillage of chemicals and how such spillages should be contained and cleaned up to ensure adequate protection of the staff and the environment but also to minimise product contamination risk;
- who should be contacted in the event there is a problem during the cleaning task;
- which cleaning records need to be completed after the task has been completed and by whom;
- who is responsible for monitoring the task to ensure it has been effective and and completes the records to that effect;
- who is responsible for verifying after the cleaning and sanitation process that the standards of hygiene are as required and that the associated records are completed appropriately; and
- who is responsible for routinely auditing the whole process.

21.4 Cleaning schedules may be designed to address daily, weekly, monthly, quarterly, six-monthly and annual tasks separately or all on one format. Cleaning may be monitored by the quality control function as part of a routine site hygiene audit, or cleaning may be monitored instead as part of production procedures by a hygiene team leader or equivalent. Verification of cleaning will occur at a prescribed frequency and by defined competent individuals such as the quality control manager or designate, and will encompass a variety of activities, including auditing of records, visual audits of premises, review of swabbing results, microbiological testing of food products, customer complaint data etc. (see 21.12).

21.5 The cleaning schedule needs to ensure effective, economic cleaning. Equipment should be installed so as to facilitate effective cleaning. The practices used will vary with the size and type of manufacturing premises and number of staff involved. Cleaning schedules should be in place to ensure that external areas are kept clean and free from rubbish. Particular attention should be paid to waste storage areas. The cleaning requirements for chill and cold stores, road vehicles and shipping and air freight containers need to be considered. Cleaning schedules need to be developed with appropriate protocols to minimise the risk of product contamination and determine the effectiveness of cleaning.

The quality control manager or designate must determine a site hygiene plan, otherwise termed cleaning and disinfection protocol, that is appropriate to the business and the products manufactured, especially in terms of high- and low-care requirements. Consideration should be given to the generation of aerosols that could contaminate nearby surfaces, packaging, ingredients or products. The control of allergenic materials and the disposal of cleaning equipment that may have come into contact with allergenic materials must also be considered.

The implementation of the cleaning schedule will only be effective if there are sufficient resources available in terms of time, trained personnel, suitable chemicals and cleaning equipment. Consideration should be given to the level of cleaning, for example routine cleaning at one level and then further deep cleaning programmes at designated intervals. Deep cleaning may need to be undertaken during periods of production shutdown and non-production, and engineering and maintenance resources may also need to be programmed where equipment dismantling is required. If equipment is dismantled, appropriate controls should be in place to ensure that machine parts are not placed directly onto the floor. Controls include, but are not limited to, tables, designated racking and so forth. A hygiene clearance/sign back into production procedure should be in place for all cleaning tasks, but additional clearance procedures may be adopted following deep cleaning activities to ensure the risk of product contamination is minimised. The cleaning schedule and associated task procedures must be validated before implementation and monitored by staff to ensure they are sufficient to deliver the food safety objectives of the manufacturing organisation. Verification activities must take place over and above monitoring to demonstrate that the hygiene management system as a whole is functioning effectively (see 21.12).

Site Hygiene Plan   21.6   The quality control manager should develop and validate a site hygiene plan and associated procedures for the site. The site hygiene plan should be reviewed at regular intervals to determine its appropriateness and effectiveness in ensuring a hygienic manufacturing site and minimising the risk of potential product contamination. The Food Standards Agency (FSA) publication *E. coli O157 – Control of cross-contamination: Guidance for food business operators and enforcement authorities* (2014) stresses the importance of not only validating the hazard analysis critical control point (HACCP) plan, but also ensuring that disinfectants are purchased and used in compliance with validated dilution levels and contact times. The guidance states that the use of disinfectants or sanitisers that meet BS EN 1276:2009, BS EN 1650:2009 + A1:2013, BS EN 13704:2002 and BS EN 13697:2015 should be considered. In order to promote efficacy they must be applied to visibly clean surfaces and be used 'strictly in accordance with the manufacturer's instructions relating to proper dilution of the chemical, the effective temperature range and the necessary contact time. Since effective chemical disinfection can only be achieved on visibly clean surfaces, a cleaning stage is required first'. Validation of cleaning and disinfection processes is especially critical if dual use of equipment and machinery for slicing, mincing or vacuum packing of raw and ready-to-eat foods forms part of the manufacturing process. Where possible, there should be designated machinery to prevent this risk from occurring. Where this is not possible, effective disinfection is a key prerequisite to ensuring food safety. Validation of cleaning methods should also assess the potential for cross-contamination by cleaning equipment. The guidance further requires that food business operators (FBOs) should 'ensure that they are using the appropriate disinfectant products by confirming with their suppliers that the products they are using meet, as a minimum, the specifications of these standards. This information may also be obtained from the label of the product, or by contacting the manufacturer directly.'[1] Records should be maintained of all validation and revalidation activities.

21.7   Risk assessment must be undertaken that is specific to the manufacturing unit to determine the level of cleaning and disinfection required. The site hygiene plan should be developed to address the following points as appropriate to the site and the type of product manufactured:

(a) The food products being manufactured will influence the requirements for those products in terms of cleaning and disinfection. Low-risk products may not require a full disinfection process to take place, especially where the product is not in direct contact with the equipment, whereas a high-risk product will require both cleaning and disinfection to be undertaken at the food premises.

[1] https://www.food.gov.uk/sites/default/files/ecoli-cross-contamination-guidance.pdf.

(b) It needs to be determined whether a clean-down process is required between different products during a production run or the cleaning process will be undertaken at the start and end of the production period (i.e. daily, per shift or between production runs).

(c) It needs to be determined who will undertake the cleaning, i.e. will there be a designated hygiene team or will the production staff undertake the cleaning at the end of production or intervals as determined?

(d) If a hygiene team is used, will they be internal staff or will cleaning be outsourced to a cleaning contractor? Cleaning contractors should be approved as per the supplier approval procedure (see Chapter 23).

(e) The budget for cleaning needs to be determined. The cost of water, equipment, labour, chemicals, energy, production down time and waste treatment should be considered.

(f) The chemical supplier should be approved as per the supplier approval procedure (see Chapter 23). The chemical supplier should be able to demonstrate compliance with the BS EN standards highlighted in 21.6 and also as applicable:
- the suitability of the chemicals for the tasks being undertaken;
- the efficiency of the chemicals especially with regard to key pathogens/microorganisms. The potential for spore formers surviving the disinfection process should also be considered;
- a knowledge of current personal health and safety at work, and environmental legislation and how it pertains to their products and the associated controls that are required to ensure compliance;
- an ability to provide material safety data sheets for the chemicals supplied;
- an ability to undertake the required level of operator training; and
- demonstrate how the chemicals should be used in practice to deliver the required level of hygiene standards.

(g) The plan should take into account the areas to be cleaned, the structure and layout of the premises and the type and condition of the surfaces for floors, walls, ceilings and doors. It should also differentiate between food and hand contact services that require a disinfection process to take place and non-hand and non-food surfaces where cleaning alone may be sufficient.

(h) The plan should also take into account the equipment type, design, purpose and the type of cleaning that is required. Some equipment can be fully cleaned without dismantling while other equipment may require full dismantling. Cleaning may involve both partial cleaning and deep cleaning. Other equipment may involve a lot of pipework and a cleaning-in-place (CIP) system may be used (see 21.11).

(i) The plan should take into account that water hardness will affect the types of chemicals that can be used.

(j) The plan needs to take into account the type of soiling on a given surface or piece of equipment and whether the cleaning task is the removal of general debris, grease and/or involves the removal of material with a high microbial loading.

(k) The plan needs to reflect the situational factors in the manufacturing unit, such as the presence of services, including water, steam, electricity and drainage.

(l) The plan needs to reflect the supervision and training needs of the staff, as well as the mechanism for assessing the effectiveness of training and the influence of the plan on the requirements for refresher training for staff.

(m) The plan needs to identify the records that need to be completed following cleaning; and

(n) An inventory and stock rotation system should be set up to ensure that the actual volume of chemical used complies with expected volumes. Any discrepancy between actual and expected usage must be fully investigated by the appropriate personnel.

21.8   It is important that the quality control manager has sufficient understanding and is competent in the development of cleaning protocols. It is critical that she/he can distinguish between the different terminology and be aware of the potential issues if an incorrect type of chemical is used. All chemicals used in food environments should be food grade, non-toxic, non-corrosive and non-perfumed, and should not cause tainting.

21.9   Care should be taken to avoid mixing of cleaning agents since their chemical nature may cause them to interact and could result in a health and safety hazard to staff working in the area.

21.10   Detergents are formulated to remove soil and dirt. Disinfection can be achieved by using:

(a) heat and/or steam as moist heat is most effective. Steam cleaners can also be used to disinfect machinery or surfaces. However, the use of heat to disinfect is costly and often impractical. As a result, chemical disinfectants are now widely used (see (b)). When using heat to disinfect equipment the efficiency is based on a relationship between temperature and time, for example equipment may be disinfected in sterilising units where equipment is immersed in water at 82 °C for 30 seconds;

(b) chemicals – the chemical disinfectant used will depend on a number of requirements, including:
   •   the amount of grease and soiling;
   •   the effectiveness of the chemical against the type of micro-organisms under consideration;
   •   the temperature and chemical contact time;
   •   the equipment and type of surfaces;
   •   water hardness;
   •   the potential for taint;
   •   the method of application; and
   •   the detergent that has been used prior to disinfection.

Disinfectants or sanitisers should meet the requirements of the relevant BS EN Standards (see 21.6) 1276:2009 or BS EN 13697:2015 (see 21.6). Liquid hand wash that has disinfectant properties should conform to the requirements of the BS EN 1499:2013 standard. This information on disinfectants and sanitisers should be available on the label of the product, or may be obtained from the supplier or manufacturer. A sanitiser is a chemical that both cleans and disinfects. A sanitiser reduces the number of cleaning steps because it has both detergent and disinfectant properties and does not require an intermediary rinse. However, sanitisers are not effective on heavy soiling and can be expensive.

Cleaning in Place 21.11 Food production systems, especially in dairies or drink manufacture, often have a number of bulk tanks, interconnecting pipework and pasteuriser systems. It would be impractical to dismantle this system for cleaning and personnel entry into bulk tanks would not be safe, so the equipment is cleaned 'in place'. Cleaning in place (CIP) allows equipment and pipework to be cleaned between processing runs. When designing a CIP system, the following factors are important:

- the cleaning requirement for each piece of equipment, run of pipework or tank/vessel;
- flow rate, temperatures and chemical concentrations;
- the mains services required for CIP cleaning, for example water temperature, automatic dosing systems and waste systems;
- the time available for cleaning;
- the programme of chemicals to be used and the temperature of the cleaning chemicals; and
- the possibilities for recycling of detergents.

A schematic plan should be available to demonstrate that the CIP system has been hygienically designed and validated (see 21.6), and a log of changes and subsequent revalidation is maintained. It is very important in the development of CIP cleaning systems that:

- the automatic chemical dilution equipment is routinely monitored to ensure it is operating correctly and the chemicals are of the required dilution;
- there is adequate separation between lines being cleaned and lines filled with products, for example double seat valves or blanks in pipework;
- there are no dead legs in the pipework that are not cleaned;
- that in-line filters do not impede the flow rate required and that they are cleaned as part of the process and inspected at a frequency defined by risk assessment;
- transfer pipes are not put into the system as a temporary measure and then are not cleaned because they are not part of the CIP system;

- the cleaning chemicals are non-foaming;
- the in-line pumps are of sufficient capacity to give the flow rate required;
- the temperatures are appropriate to ensure effective cleaning;
- the cleaning times are validated to ensure that there is sufficient contact time for the chemicals with surfaces;
- the levels of chemical in stock can be determined to identify if the correct amount of chemical has been used;
- the rinse water is checked to ensure that all chemical has been removed before the system is put back into production, for example pH monitoring;
- the location and mesh size of filters is determined as well as the monitoring activities associated with the filters;
- spray-ball or rotary cleaning head systems in tanks are checked to ensure that all areas of the tank receive a coverage of chemical and there is an adequate spray pattern; and
- all flexible hoses are cleaned as part of the CIP routine. Flexible hoses should also be capped and stored in a designated area when not in use.

CIP systems must not only be validated, but also monitored and verified to ensure continued effectiveness. Activities that should be undertaken include analysis of rinse water and/or first product through the lines (which is sent to waste until approved as suitable for packing), usage data for cleaning chemicals and ATP bioluminescence testing.

Efficiency of Cleaning

21.12 The efficiency of cleaning and sanitisation should be checked and recorded at routine intervals by the quality control manager or designate. The frequency of checks should be based on risk assessment. The hygiene standards required should be defined. Mechanisms of monitoring and verification of the effectiveness of cleaning include:

- visual inspection;
- contact microbiological swabs;
- ATP bioluminescence techniques;
- food material and product testing, where risk assessment has identified the potential for chemical residues to be present in food if cleaning procedures are not suitably followed;
- microbiological testing of part-processed and finished products; and
- microbiological and chemical checks of rinse water.

The monitoring and verification of cleaning activities should be formally documented, records maintained, any trends analysed and, where required, corrective action implemented to improve cleaning and sanitisation practices. Follow-up inspections should be undertaken to ensure that the corrective action has been implemented and is effective. These records should be maintained as part of the due diligence system.

Biofilms

21.13 The formation of biofilms is a major concern in food manufacturing environments. Biofilms are caused by micro-organisms that adhere to each other and also to a surface. If biofilms form then this will reduce the efficacy of disinfection systems, especially as they can provide harbourage for pathogenic and spoilage organisms such as *Salmonella* spp. and *Listeria monocytogenes*. The type of biofilms formed will depend on the environment within the manufacturing unit in terms of the products manufactured and processed, and thus the nutrients and pH within the environment, and temperature. The risk of biofilm formation is greater in areas of the factory which are cleaned and/or dismantled at a lower frequency, e.g. in-line filters or pipework joints and seals, and there is potential for sloughing from the biofilm into liquid if there is a change in flow rates and pressure. It is critical in the manufacturing environment to be aware of the potential for biofilm formation and to develop cleaning programmes and associated monitoring and verification that minimise the likelihood of biofilms occurring.

Equipment and Storage

21.14 There must be controls in place to ensure safe and secure storage of cleaning and sanitisation chemicals when they are not in use. The quality control or hygiene manager, as appropriate, should develop procedures to ensure that:

- chemicals are kept in secure, closed, labelled containers and used according to manufacturers' instructions;
- chemical stores are secure, locked and bunded, and away from food areas;
- stock is rotated correctly and chemicals are used within their duration marking;
- food containers are not used for storage of cleaning chemicals;
- health and safety data are available and reliable for all chemicals;
- cleaning chemicals for food-processing areas and for toilet areas where these are non-food grade are stored separately;
- correct protective clothing is used; and
- spillage procedures are developed and implemented.

21.15 Appropriate cleaning equipment must be used. It should be fit for purpose and not be a source of contamination in itself. Consideration should therefore be given to using non-wooden equipment and/or equipment that is so coloured that a foreign object can be easily identified in the finished product, for example brush bristles. Cleaning equipment that is used in designated areas should be colour coded to prevent its use in another area, for example external equipment, internal machines, floor cleaning, internal factory, toilets and welfare areas. This is especially important in manufacturing units where there are designated high- and low-care areas. A colour-coded site plan may prove effective in communicating where equipment can be used. The use of mops, especially string mops, should be risk assessed to determine that they can be effectively cleaning and disinfected after use.

Cloths should be adequately controlled and be single issue where deemed appropriate within the hygiene risk assessment. Surfaces should be allowed to air-dry, but if the task procedure requires that they must be wiped after cleaning, this must be with a single-use cloth, a sanitising wipe or blue paper towel that is appropriately disposed of to prevent contamination. If an item is found to be incomplete, for example dustpan and plastic handle broom, then the brittle material control procedure should be implemented (see 19.36–19.42). If disposable single-issue gloves are used by personnel during cleaning, they should be adequately controlled. The equipment should be cleaned and stored so as to minimise the potential for product contamination. Equipment should be allowed to air dry in designated racks.

22 INFESTATION CONTROL

The protection of food against contamination from pests is a key prerequisite within a good manufacturing practice (GMP) system. Preventive methods include staff awareness training, adequate proofing, an effective pest control programme, minimising harbourage and potential food sources, and suitable control of incoming ingredients and materials that could support pest infestation. The major emphasis, however, must always be prevention.

22.1 The **Prevention of Damage by Pests Act 1949** requires an occupier of any land or buildings to notify the local authority of rodent infestation. It also requires all local authorities to take steps to ensure that their district is free from rats or mice.

22.2 The **Food Safety Act 1990** made it an offence to sell food that is unfit for human consumption or contains foreign bodies, for example pests, parts of pests or droppings. **EU Regulation (EC) No. 852/2004** on the hygiene of foodstuffs requires that 'The layout, design, construction, siting and size of food premises are to … permit good food hygiene practices, including protection against contamination and, in particular, pest control'. To do this requires measures including:

- proofing of entrances and other access points;
- insect screens on open windows;
- electronic fly killers (EFKs), otherwise known as insectocutors;
- minimising harbourage, especially good stock rotation of dry goods;
- regular surveys by competent pest control contractors; and
- the use of appropriate baits and chemical pest control measures.

22.3 The principles of pest control are to limit food supply, access or movement of pests, and potential harbourage. There are two types of pest control: physical and chemical. Physical control methods are preferable as the pest is caught, cannot further contaminate the food and there is no risk of chemical contamination of food. Physical methods also include proofing of the building. Physical methods are not always completely effective, and a combined approach with chemicals may need to be used. A major concern with chemical control is that the pest may not be killed at the point of contact, for example ingestion, and could die later and possibly contaminate the food.

22.4 Incoming ingredients and materials should, as appropriate, be thoroughly inspected for signs of pest infestation as part of the quality control programme. All materials should be stored (see 22.10) to minimise the risk of infestation.

Food & Drink – Good Manufacturing Practice: A Guide to its Responsible Management, Seventh Edition.
The Institute of Food Science & Technology Trust Fund.
© 2018 John Wiley & Sons Ltd. Published 2018 by John Wiley & Sons Ltd.

22.5 The most effective contribution towards infestation control is maintaining good housekeeping standards, for example controlling accumulations of food and paper debris, keeping gangways and passages clear, removing redundant equipment and materials from the manufacturing areas, and ensuring good stock rotation. External housekeeping is as important as internal housekeeping. External housekeeping includes:

- keeping vegetation short and not allowing vegetation to grow close to buildings;
- managing waste, ensuring all external waste areas are enclosed and they do not encourage pests nor provide harbourage;
- ensuring that leaking taps are not providing a source of water for pests to drink;
- ensuring waste building materials, redundant equipment, pallets or wooden boxes are not allowed to accumulate close to manufacturing units; and
- ensuring attention is given to other activities in the vicinity that could attract pests, for example adjoining farming activities, food manufacture or storage, or waste handling.

22.6 All premises should either have personnel trained in pest control or employ a professional infestation control organisation for regular inspection, advice and treatment to deter and destroy infestation. An outside contractor should be accompanied on his/her visit by an appropriate staff member. It is important at the end of each visit to have a closing meeting with the contractor to discuss key points such as the current level of pest activity, contractor recommendations on proofing or general housekeeping, areas where the contractor has not been able to gain access to examine bait points and whether any actions recommended in previous visits have been addressed appropriately by the manufacturing organisation or if further action is required and if the documentation in the pest control file is up to date.

22.7 Site inspections should be regular and reports of each inspection should be kept on file to be available for future reference if required (see 22.16).

22.8 Care should be taken to ensure that bait spillages cannot pose a risk to any foodstuffs being manufactured or stored in the vicinity. Wherever practicable rodent baits should be based on fatty or waxy substrates. Risk assessment of individual locations should be undertaken to determine whether loose or solid bait materials will be used. The risk assessment should also assess whether only non-toxic baits should be used in internal areas. Private standards differ in their prescriptions on this point and their requirements should be determined in order to ensure compliance and thus approval. Standards, such as the British Retail Consortium (BRC) Global Standard for Food Issue 7, prescribe that only non-toxic baits can be used within production or storage areas where open product is present except when treating an active infestation (see 22.16).

22.9 No substances or application methods should be used in infestation control that have not been approved under the **Control of Pesticides Regulations 1986 as amended 1997**. Due attention should be given to risks of cross-contamination, for example by spillage of grain bait, inadequate cleaning after chemical application to control pests and dispersal of tracking dust. Care should be taken to protect all materials, product, packaging, utensils and surfaces in contact with food from residual contamination by pest control substances.

22.10 Goods and equipment should always be stored at least 50 cm from adjacent walls to facilitate cleaning and inspection for rodents and insects. Drains and gutters should be fitted with screens and traps to prevent pest entry. Attention should also be paid to drains and drain pipes that could allow pest ingress.

22.11 Except where there are risks of dust explosions, all production areas should be supplied with electrical insect killing 'knock-down' devices. These should be sited in areas of minimum light intensity for best effect, but should not be sited directly above any tanks, hoppers, conveyors, processing equipment, filling machines or open food.

22.12 Depending on insect type and the products being produced, pheromone traps may be used, where appropriate. Insect 'knock-down' devices should be fitted with suitable catch trays that should regularly be inspected and emptied when necessary (see 22.16). These devices should be left switched on at all times even after production has finished and the premises vacated; this is particularly important during the summer months. The ultraviolet (UV) light tubes on these units should be shatterproof in design and replaced at least annually. The glass tubes should be suitably controlled under the site's brittle material control procedures. The annual bulb change of the UV light should be recorded, plus any damage or breakage to the light bulb and the actions taken (see Chapter 19, 19.36–19.42).

22.13 Domestic animals, for example cats and dogs, should be excluded from manufacturing and storage areas. This is mainly accomplished by keeping all doorways and other entrances closed. Staff should never feed or otherwise encourage stray animals. Waste skips and waste storage areas should be adequately controlled to prevent attracting feral cats, especially on meat-processing sites.

22.14 Birds and insects must be excluded from all production and storage areas. To do this effectively, all apertures in the roof or its eaves, or the walls should be identified and either closed off or suitably screened. Drains and guttering should also be fitted with screens and traps to prevent pest access. Doors and windows similarly should be suitably protected, for example by use of air curtains, strip curtains and netting (see 19.11).

22.15 Where there are already birds in the confines of the premises, they, and their nests, must be removed. In the UK there are legal requirements under the **Wildlife and Countryside Act 1981**, elsewhere due regard to the requirements of the various bird protection agencies must be observed.

Pest Control Programme

22.16 The pest control programme should address the following steps:

(a) *Assessment:* The pest control programme can be undertaken internally by suitably trained staff, or a contractor could be engaged to undertake either part of or the entire pest control programme. If an organisation is using a pest control contractor, they must determine the suitability of the contractor and should assess criteria including the following:

- if the manufacturing organisation's customers have an approved list, then one of these approved organisations must be used to deliver the pest control programme;
- the proximity of the pest control contractor to the manufacturing site, the resources available and the response time in the event of a problem, information such as number of employees and holiday cover, and ease of communication, for example mobile telephone numbers and weekend cover;
- whether the pest control contractor can demonstrate technical competence, that is, in the UK is a member of a recognised association such as the British Pest Control Association (BPCA; http://www.bpca.org.uk) or National Pest Technicians Association (NPTA; http://www.npta.org.uk), and the individual operators have formally recognised training and/or qualifications and if required have been on refresher courses to keep their knowledge and skills up to date;
- the experience of the pest control contractors in the food industry sector and the specific types of food produced and whether they can supply references from current clients;
- the type of contract the pest control operator provides and whether it meets internal/external requirements in terms of types of pests covered and the frequency of visits, that is, the ability of the contractor to undertake a complete pest risk assessment, undertake routinely the pest survey required and provide a clear report of their recommendations and actions required. The contract needs to address the frequency of the contractor's field biologist quality assurance surveys as well as the number of routine pest control visits; and
- evidence of the financial viability of the contractors and their current level of insurance cover with regard to product, public and employer's liability. A copy of current insurance certificates for the pest control contractor should be retained within the pest control manual.

(b) *Protocol:* The pest control contractor should provide a pest control manual that includes the following:

- a health and safety risk assessment for the pest control contractor employees identifying any health and safety concerns. Material safety data sheets must be retained along with Control of Substances Hazardous to Health (COSHH) assessments for the insecticides or rodenticides and other substances that the pest control contractor could use to demonstrate compliance with the **COSHH legislation** (or the equivalent outside the UK). These data should also be available for non-toxic baits if they are used as well as toxic materials. The pest control contractor in the UK must demonstrate compliance with the **Control of Pesticides Regulations 1986 as amended in 1997** and the **EU Biocides Regulation No. 528/2012**;

- a contract for the pest control service, based on risk assessment, identifying the minimum number of routine inspections, emergency call-out requirements, quality assurance inspections and the types of pests controlled, that is, the scope of the pest control programme. The risk assessment should also consider external risks based on location, incoming and in-process materials and products produced, for example stored product insects will be a factor with dry materials, and the combination of physical and chemical pest control methods that will be used. The pest risk assessment may also identify the number of field biologist visits required per annum. The initial risk assessment must determine the locations for toxic and non-toxic rodenticide. Toxic bait should not be used in areas where materials and products are subject to open storage, processing or other associated manufacturing areas unless a risk assessment has deemed that the risk level is such that the potential contamination hazard is constantly controlled to an acceptable level. If toxic baits are used as a result of an active infestation, they must be adequately controlled to prevent contamination;

- a dated bait plan, pheromone trap and/or EFK plan identifying the premises or site layout and where the baits or equipment are located. The bait points/EFK should be individually numbered and identified on the plan, and this number should correspond to the actual number on the bait box or item of equipment. The bait boxes should be robust and tamper proof, that is, hard plastic rather than paper box design, and also be secured to prevent loss or accidental damage. In the event that a bait box is found to be missing, the quality control manager or designate must undertake an investigation to identify the whereabouts and/or potential risk to the product and action must be taken as appropriate. The bait plan should also include areas of roof void and any baiting that has been undertaken. It should be signed off by both the appropriate member of staff from the manufacturing site and the appropriate person from the pest control contractor company. The bait plan should then be reviewed and re-signed and dated on an annual basis to demonstrate that verification activities have been undertaken and that there have been no changes. In the event of changes, the bait plan will need to be reissued and authorised by the two personnel as previously described;

- a checklist for bait points and EFK, which identifies the status of each point on each inspection for easy reference and analysis of trends. The coding is usually $R = rat$, $M = mice$ or $I = insects$. The EFK should be individually identified. The trays should be routinely inspected and data collated, including the numbers of flies, the types of flies and any trends noted. Action must be taken as appropriate;
- inspection reports, which should be maintained by the pest control contractor for each inspection identifying the pest control status, any evidence of pests, proofing or poor housekeeping and any recommendations. The reports should be numbered for easy cross-referencing. Any required corrective actions need to be signed off by the pest control company to demonstrate completion. The reports should be routinely analysed by the quality control manager for trends;
- current membership certificates for the BPCA or equivalent;
- current insurance certificates; and
- training certificates, which should be retained to demonstrate the competence of the operators undertaking pest control and those undertaking quality assurance activities.

(c) *Implementation:* The pest control contractor needs to implement the agreed programme, and personnel on the site need to be vigilant to ensure that they continue to check for evidence of pests. The manufacturing organisation should never effectively delegate responsibility for pest control to the contractor, partly because it is site personnel who are most likely to see evidence of pests or the pests themselves. Staff must be made aware that they must report evidence immediately to their line managers so that the manufacturer is actively managing its pest control system and an emergency visit can be arranged with the pest control contractor. The pest control contractor should always be accompanied by a competent member of staff to ensure any issues identified are suitably communicated during the routine visit or an emergency visit if required;

(d) *Evaluation:* The effectiveness of infestation controls/the pest control programme and the compliance of the pest control contractor with the requirements of the contract both need to be evaluated. An audit of the pest control procedure needs to be carried out as part of the internal audit process (see Chapter 11). The management control of the design, maintenance and proofing of buildings is critical to the pest control programme. It is not enough to have a pest control contractor visiting the site if the doors are left open or do not fit close enough to the floor, fly screens are broken or drains, roof spaces, piping and ducting allow access. The effectiveness of the premises hygiene protocols should also be assessed and their potential impact on the pest control programme. If potential food sources are not adequately controlled, then this will reduce the effectiveness of the pest control programme. Trend analysis is an important element of this evaluation.

22.17   The quality control manager and their nominated deputy must be responsible for the implementation of the pest control programme. They should be responsible for undertaking a review at regular intervals with the pest control contractor to ensure that the required level of service is maintained, for example the number of visits complies with the contract and that the programme provides effective control, there is a hazard data sheet in the file for each chemical control used and so forth.

23   PURCHASING

Principle

Suppliers should be selected on their ability to supply product or service that meets defined requirements. The exact nature of the supplier approval and performance monitoring undertaken will depend on the effect of the purchased product or service on the finished product. The control of incoming products and services is a key prerequisite within a good manufacturing practice (GMP) system.

General

23.1   The management team should develop a supplier quality assurance and performance monitoring procedure that defines the criteria for selection, approval, review and ongoing approval of suppliers and the materials and services they supply to ensure that purchased products and services consistently meet the manufacturing organisation's requirements. The extent, scope and nature of the procedure will relate to the food products produced and the interface of materials and services purchased with the ability of the manufacturer to produce safe and legal product(s) to the required food safety, integrity and quality standards.

Supplier Quality Assurance

23.2   The quality control manager should have an advisory role in the choice of suppliers of raw materials, packaging, equipment and services. This should include determining if the supplier possesses certification to third-party assurance technical/management standards or internal company standards, certificates of analysis/conformity and/or in-house checks. She/he should liaise with actual or potential suppliers in agreeing the relevant material or service specifications, including the assessment of their ability to consistently meet those specifications. The quality control manager should have authority to exclude any actual or potential supplier of material, packaging, equipment or services on the grounds of inability to meet the relevant specification or of unsatisfactory performance or inadequate food safety, integrity or quality control resulting in unreliability of product or service. Assessment of the current status of the supplier's food safety management system (FSMS), hazard analysis critical control point (HACCP) plan, food integrity management system (FIMS) and their quality management system (QMS) should also be undertaken as relevant to the materials and services supplied. There should be continuous surveillance of supplier performance and internal review of the outputs of that surveillance at designated intervals. The intervals determined for monitoring and/or verification checks should reflect the supplier's history of conformance with the food manufacturer's requirements. An approved supplier list should be developed, maintained and routinely reviewed (see 23.7), and this activity should be defined in the supplier quality assurance and performance monitoring procedure.

Food & Drink – Good Manufacturing Practice: A Guide to its Responsible Management, Seventh Edition.
The Institute of Food Science & Technology Trust Fund.
© 2018 John Wiley & Sons Ltd. Published 2018 by John Wiley & Sons Ltd.

23.3 The purchasing manager should have authority to place orders with suppliers on the approved suppliers list for raw material lots and services that conform to the relevant specifications. For those raw materials customarily sold on 'buying sample', the quality control manager should examine and report on the buying sample, and the purchasing manager should not have authority to disregard adverse findings. Where a delivery of a purchased raw material is found not to conform to the relevant specification or the buying sample, the purchasing manager should initiate and pursue appropriate action with the supplier as defined in the supplier quality assurance and performance monitoring procedure.

23.4 The materials, packaging, equipment and services that should be within the scope of the supplier assurance and performance monitoring procedure include:

(a) raw material and ingredient suppliers, both direct suppliers, which could include agents and distributors, and, if required, primary producers;
(b) packaging and food contact materials suppliers, including primary and secondary packaging suppliers;
(c) water suppliers;
(d) equipment suppliers;
(e) contract services such as maintenance and servicing, calibration, contract cleaning and retail tray washing services, product testing and laboratory services, transport and distribution services, training providers, contract labour and agency staff, pest control, laundry and catering services, hygiene and waste management and disposal services;
(f) subcontracted or outsourced manufacturing and storage operations; and
(g) processing aids.

23.5 Ingredient and direct food contact materials (including equipment) will require specific risk or HACCP-based assessment to determine the areas that are critical to ensure food safety, legality and quality. This assessment will include, but is not limited to, the food safety hazards addressed in Chapters 3, 8, 24 and others. Any exceptions to standard procurement and approval procedures and what actions will be taken should be identified on a case-by-case basis, for example where raw material suppliers are prescribed by retail customers, where products are purchased from cooperatives, marketing organisations or agents and where emergency procedures have meant that an alternative supplier is required at short notice to ensure continuity of supply and the timescale does not afford full compliance with the supplier approval and performance monitoring procedure. Factory trials and product development ingredients may also be deemed as exceptions; however, adequate procedures must be in place to ensure that food safety, legality and quality are fully maintained.

23.6 Supplier approval and performance monitoring should be developed on a risk basis and can include, but are not limited to, the following:

- approval of pre-supply samples, including product performance standards;
- determination of the volume of material supplied and the inherent food safety hazards associated with the material;
- assessment of supplier history and their level of historic compliance with requirements;
- pre-audit of suppliers' premises to ensure they meet the required standards, especially in terms of hygiene and the implementation of prerequisite programmes (PRPs) as well as the standards defined within this guide or designated system or customer standard;
- pre-audit of supplier QMS, FSMS or FIMS from either documentary evidence or completion of a supplier questionnaire. In some cases, both may be necessary. The timescale for reissuing of questionnaires should be determined in the supplier approval and performance monitoring procedure and be based on risk assessment. Suppliers should be required to advise the manufacturer if there is a change to the responses they have made to the 'live' questionnaire, for example a loss of certification, product withdrawal or recall, so that the manufacturer can continue to review performance on the basis of all available facts;
- monitoring of their performance during a 'trial' period to determine the supplier's ability to deliver products and services that comply with specified requirements;
- monitoring of their performance over time in complying with the appropriate specifications;
- demonstration by means of a current certificate that the company is certified to a recognised management system standard, for example BS EN ISO 22000:2005 or the British Retail Consortium (BRC) Global Standard for Food Safety;
- certification to an appropriate QMS or FSMS standard for distribution, transport, forwarding, shipping and air transport companies, for example ISO 9000 or relevant BRC standard;
- certificates of analysis or conformity with each delivery;
- development of material sampling and verification procedures to check the veracity of supplier information and pre-supply samples on an ongoing basis. This could include pesticide residue analysis, heavy metal analysis and microbiological analysis. These sampling and verification activities may be undertaken by the manufacturer or supplier as formally agreed;
- review in the event of the identification of an emerging food safety hazard as a result of new scientific or technical information or a food safety incident that has been identified within the food supply chain;
- follow-up of instances of non-conformity of materials and/or service;
- history of supplier in terms of food integrity and more specifically risk of food fraud, a food defence issue or associated illegal behaviour (see Chapter 5); and

- supplier audits and/or inspection checks, where practical, periodically made on suppliers based on risk assessment.

This information is crucial to ensure that suppliers behave with integrity and provide materials and/or services of the correct and consistent quality and that they have addressed any potential food safety, integrity, quality or health and safety issues specific to the materials, equipment and services they supply. This information also provides input into the management review process, ensuring that any non-conformity is reviewed and then appropriate action taken if required (see Chapter 11).

Supplier audits must be undertaken by competent individuals. An associated audit report must document the audit findings and where necessary any agreed preventive or corrective action. Records showing how the competency of company auditors and/or third-party auditors has been determined, validated and routinely verified should be retained for a prescribed time period by the quality control manager. The completion of preventive and corrective actions by the supplier, when requested, must be verified to ensure that they have been completed in a timely fashion and that they have been effective.

23.7   The supplier approval and performance monitoring procedure should address the actions to be taken in the event of material, packaging, equipment, finished product and/or service non-conformity, including supplier delisting procedures. A formal annual supplier review should be undertaken where each supplier's level of performance is reviewed and any actions required are determined and implemented. Questionnaires should be reviewed and updated at designated intervals to ensure that they continue to reflect supplier standards and third-party certification status. Responsibilities and timescales for completion should be developed for any resultant actions, which should be followed up at the appropriate time interval to ensure that they have been implemented and have been effective.

23.8   The frequency of testing or verification of purchased materials, packaging, equipment and/or service should be determined by the quality control manager. Materials should be inspected on receipt to ensure compliance with specifications and that traceability has been maintained from the original source. On-receipt certificates of analysis, certificates of conformity and pre-acceptance testing (positive release testing) may be required depending on the initial risk assessment undertaken. Packaging that does not undergo quality checks on receipt should be monitored as part of the production quality control checks. Material and supplier risk assessments must be reviewed on a minimum of an annual basis and whenever non-conformity is identified. Appropriate action must be undertaken to ensure that any changes to procedures and specifications are validated and implemented, and are effective.

23.9 As previously mentioned in 23.4, the supplier approval and performance monitoring procedure also needs to address the range of outside services that could be purchased under contract. These include:

- building repair and maintenance;
- laundering;
- waste disposal;
- specialist engineering maintenance;
- warehousing;
- computer services; and
- advisory services such as business consultants, technical consultants, marketing consultants and legal consultants.

23.10 The contract giver should fully satisfy themselves that a contract acceptor has the appropriate knowledge, skill, facilities, equipment, staff and so on to be able to provide the service in such a way as to contribute to the objectives of GMP. Except in the case of purely advisory service, if the work is carried out away from the contract giver's premises, the contract giver should have and should exercise the right to visit and inspect any location where the work is performed.

23.11 A contract acceptor should not pass on to a third party any of the work commissioned by the contract giver except with the prior knowledge and consent of the latter.

23.12 In the absence of specific authorisation to the contrary by the contract giver, the contract acceptor should treat as confidential any direct or incidental knowledge gained about the contract giver's business.

23.13 Arrangements between a contract giver and a contract acceptor should be in writing and should include adequate indication of the agreed respective responsibilities of the parties, and of any special hazards of which either party is aware, and record the authority for the contract acceptor to conduct the work.

Illicit Behaviour   23.14 In agreeing to purchase materials or services from a supplier the manufacturer should consider the potential for illicit behaviour. The scope for illicit behaviour is vast within the food sector so the organisation needs to undertake a focused risk assessment. The scope needs to be defined and should include the materials or products that pose the most risk, the potential types of threats and perpetrators, and how the risk can be effectively mitigated (see Chapters 6 and 7).

23.15 Potential issues that can arise are detailed in Chapter 5, such as product adulteration, substitution, counterfeiting and misrepresentation among others.

24   PACKAGING MATERIALS

Principle

Packaging can be described as the materials used for the protection, preservation and presentation of food products. Product packaging should comply with relevant legislation and conform to agreed specifications. It should be stored in a designated area and in such a way as to minimise contamination or damage. Packaging procedures are a key prerequisite to ensure good manufacturing practice (GMP).

General

24.1   The principles of GMP with regard to packaging have been reviewed in Chapter 10 (10.24–10.35).

Storage

24.2   Packaging materials should be stored in a designated area separate from raw materials and finished products. If open containers are not stored in external packaging, then, where possible, they should be inverted to minimise the risk of product contamination. Packaging liners should be coloured to minimise the risk of product contamination. Stored packing should be shrink-wrapped together to provide security and stability on the pallet and to minimise the risk of any potential foreign body contamination.

24.3   Packaging materials consist of:

- primary or sales packaging;
- secondary or grouped packaging:
- tertiary or transport packaging; and
- transit packaging.

The main materials that are used for primary, secondary, tertiary and transit packaging include paper, glass, aluminium, steel, plastic and wood. **Commission Regulation (EU) No. 10/2011 of 14 January 2011** on plastic materials and articles intended to come into contact with food establishes specific requirements for the manufacture and marketing of plastic materials and articles:

(a) intended to come into contact with food;
(b) already in contact with food; or
(c) which can reasonably be expected to come into contact with food.

Commission Regulation (EU) 2017/752 of 28 April 2017 amends and corrects **Regulation (EU) No. 10/2011. Commission Regulation (EU) 2016/1416 of 24 August 2016** also amends and corrects **Regulation (EU) No. 10/2011** and comes into force in September 2018. The legislation sets out both specific migration limits (SMLs) and overall migration limits (OMLs) for specific substances and the regulations should be consulted directly for more details.

Food & Drink – Good Manufacturing Practice: A Guide to its Responsible Management, Seventh Edition.
The Institute of Food Science & Technology Trust Fund.
© 2018 John Wiley & Sons Ltd. Published 2018 by John Wiley & Sons Ltd.

Packaging materials should be suitable for the food they contain and should be inert during packing, processing, storage and distribution and in preparation by the consumer for consumption, for example microwaving or cooking in the pack. Consideration should also be given to the pH in the case of low-acid and high-acid foods and the fat content of the food and whether this will impact on packaging design and suitability. Packaging specifications should be developed and approved for all packaging materials. Packaging specifications may be provided by the customer, especially where own-label branded products are being supplied or are originated by the manufacturer. Packaging specifications should contain the following information as appropriate:

- supplier name and address and manufacturing site name and address if these are different;
- material composition;
- packaging dimensions, including thickness and gauge;
- specific storage and handling requirements, for example temperature and humidity, single or double stacking;
- material suitability and confirmation of adequate migration testing results and compliance with legislation (see 19.22);
- information on adhesives used if appropriate (see 19.22);
- artwork details, including size of lettering, bar code details, marketing and farm assurance logos and colour standards; and
- promotional information if there is both standard and promotional packaging utilised for the same product.

24.4   Some transit packaging, such as field trays, field bins or retail plastic trays, is reusable/returnable. These trays are used, for example, as a display outer for fresh products, including fresh produce and meat. These trays are sometimes referenced as returnable transit packaging (RTP). While all foods are within primary packaging, it is important that these trays are of a consistently high level of hygiene and cleanliness. It is also important to ensure that all previous tray/box labels are removed before use to ensure that identity and traceability are maintained. Risk assessment should be undertaken with loose food, for example fruits and vegetables that are packed in a tray liner to determine whether the hygiene of the trays could detrimentally affect the food safety and quality of the product. If this is deemed a risk, then suitable controls need to be put in place and verification activities undertaken to determine their continued effectiveness.

24.5   Tray washing between retail outlets and manufacturing plants is often undertaken by third-party companies. These companies should be subject to the supplier approval and performance monitoring procedure (see Chapter 23). The quality control manager should develop procedures to ensure that retail tray monitoring procedures are implemented to ensure that hygiene levels are maintained and labels have been removed and that appropriate action is taken in the event of non-compliance.

24.6   Particular attention should be paid to the temperature of the tray wash tank (ideally 75–80 °C) and the rinse tank (ideally 80–85 °C). The Fresh Produce Consortium guide *Hygiene Procedures – A Code of Practice for Returnable Transit Packaging at Traywash Units* (March 1998) suggests that, based on a maximum number of 1200 trays per hour, trays are exposed to temperatures for the following times:

- wash tank   30 seconds
- detergent rinse   7 seconds
- final rinse   13 seconds

These temperatures and times should be routinely monitored.

24.7   Retail trays, where possible, should not be stored outside. They should be shrink-wrapped to minimise contamination.

24.8   Packaging should be stored according to manufacturer's instructions to minimise damage and possible effects on packaging integrity. For example, reels of laminate packaging need to be stored at the correct conditions and in a way to minimise damage so that the laminate remains intact and does not allow air ingress.

24.9   Heat-preserved products require the maintenance of packaging integrity to ensure control of product safety and quality characteristics. The quality control manager should develop procedures to monitor packaging integrity and packaging seal quality where this is required, for example in canning, aseptic packing, vacuum packing and gas flushing operations. This will include the need to develop appropriate sampling plans. Protocols for accelerated product durability testing should be developed where appropriate.

24.10   Packaging should be traceable on a batch basis, and the quality control manager or designate should develop a packaging traceability procedure to ensure that control is maintained. Food safety hazards associated with packaging have been discussed (see 9.2, 9.14–9.18 and 19.22). Packaging suppliers should be monitored and approved as per the supplier approval and performance monitoring procedure (see Chapter 23).

24.11   The **EC Directive on Packaging Waste (94/62/EC)** aims to reduce the volume of packaging waste going to landfill sites by setting targets for the recovery and recycling of packaging waste. The UK legislation that implements the EC Directive is the **Producer Responsibility Obligations (Packaging Waste) Regulations 2007 as amended and 2016 amendments**[1] (first came into effect in 1997), which cover aspects of recycling and recovery. Guidance on current of requirements and obligations can be found at https://www.gov.uk/guidance/packaging-producer-responsibilities.

---

[1] https://www.legislation.gov.uk/ukdsi/2016/9780111142431.

The Packaging (Essential Requirements) Regulations 2015[2] (first came into effect in 2003) consolidates and revokes all earlier Regulations that relate to the essential requirements for packaging that is within the heavy metal concentration limits. The essential requirements are:

(i)   Packaging must be so manufactured that the packaging volume and weight are limited to the minimum adequate amount to maintain the necessary level of safety, hygiene and acceptance for the packed product and for the consumer.

(ii) Packaging must be designed, produced and commercialised in such a way as to permit its reuse or recovery, including recycling, and to minimise its impact on the environment when packaging waste or residues from packaging waste management operations are disposed of.

(iii) Packaging must be so manufactured that the presence of noxious and other hazardous substances and materials as constituents of the packaging material or of any of the packaging components is minimised with regard to their presence in emissions, ash or leachate when packaging or residues from management operations or packaging waste are incinerated or landfilled.

The aggregate heavy metal limits apply to cadmium, mercury, lead and hexavalent chromium in packaging or packaging components and require that the total by weight of such metals should not exceed 100 ppm (subject to some exemptions for some products).

[2] http://www.legislation.gov.uk/uksi/2015/1640/contents/made.

25   SMART PACKAGING

Principle

Smart packaging can be described as packaging systems that have enhanced functionality when compared to standard types of packaging. Smart packaging can be divided into two types: active packaging and intelligent packaging. The advantages of using such packaging systems is that they can either provide functionality such as moisture control or shelf-life extension or be used to communicate information about the product in terms of the product's food safety, quality or integrity status.

General

25.1   The terms 'smart packaging', 'intelligent packaging' and 'active packaging' have gained widespread use in the food manufacturing sector and are considered in this chapter in terms of their function beyond inert or passive containment of food and their ability to extend product shelf-life and to communicate information which has historically been done through the use of product duration codes such as 'use by' or 'best before' (see 37.8–37.9). Principles of GMP with regard to packaging have been reviewed in Chapters 10 (10.24–10.35), 24 and 37. **Regulation (EC) No. 1935/2004** on materials and articles in contact with food states that such materials and articles should undergo a safety assessment prior to their authorisation. **Regulation (EC) No. 450/2009** on active and intelligent materials and articles intended to come into contact with food details legislative requirements that need to be met with regard to active and intelligent materials and articles. **Regulation (EC) No. 450/2009** provides the following definitions:

- *active materials and articles* means materials and articles that are intended to extend the shelf life or to maintain or improve the condition of packaged food; they are designed to deliberately incorporate components that would release or absorb substances into or from the packaged food or the environment surrounding the food;
- *intelligent materials and articles* means materials and articles which monitor the condition of packaged food or the environment surrounding the food;
- *functional barrier* means a barrier consisting of one or more layers of food contact materials which ensures that the finished material or article complies with **Article 3 of Regulation (EC) No. 1935/2004** and with **Regulation (EC) No. 450/2009**;
- *releasing active materials and articles* are those active materials and articles designed to deliberately incorporate components that would release substances into or onto the packaged food or the environment surrounding the food;
- *released active substances* are those substances intended to be released from releasing active materials and articles into or onto the packaged food or the environment surrounding the food and fulfilling a purpose in the food.

Food & Drink – Good Manufacturing Practice: A Guide to its Responsible Management, Seventh Edition.
The Institute of Food Science & Technology Trust Fund.
© 2018 John Wiley & Sons Ltd. Published 2018 by John Wiley & Sons Ltd.

25.2 Appropriate documentation must be maintained by the manufacturer that demonstrates that the active and intelligent materials being used comply with the requirements of **Regulation (EC) No. 450/2009**. Documentation should include information on the suitability and effectiveness of the active or intelligent material or article, the conditions and results of testing, calculations or other form of analysis, and evidence on the safety or reasoning used that can demonstrate legislative compliance (see Annex II of the legislation). To enable consumers to identify non-edible elements of a packed food product the materials must be labelled or marked as to their not being edible.

25.3 Active packaging is a packaging system with functionality whereby the product, the packaging and the external environment interact to modify the condition of the packed material with features such as moisture control, absorption of liquid or oxygen, or the release of preservatives and other forms of shelf-life extension. Active materials may deliberately incorporate substances that are intended to be added into food substance and should therefore comply with all legislative requirements associated with food additives. Alternatively, additives, including enzymes, may also be immobilised on the packaging medium and these too should comply with all legal requirements associated with additives (see Chapter 52). Active packaging can include active films that contain phenolic compounds to improve the oxidative stability of beef by reducing lipid oxidation and incorporation of preservatives into edible films and coatings.

25.4 Intelligent packaging contains features that monitor the condition of the packed material and as a result provide information on changes to the product and/or the product's status. They should not release packaging system constituents into the food. Intelligent packaging systems rather than being in direct contact with food may be located on the outer surface of the food product and thus are separated from the food itself with a functional barrier. This functional barrier will prevent substances from migrating into the food. Whilst the use of electronic data interchange (EDI) technology, especially barcode development, quick response codes (QR), radio frequency identification (RFID) and capturing data with regard to expiry dates has been reviewed in Chapter 16, other technologies are also being used to display information on quality, improve product safety and act as a deterrent for food crime such as counterfeiting. Chemicals or biosensors that form part of the packaging can convey information within the manufacturing setting or to the consumer via monitoring the presence or absence or level of a substance, or a chemical reaction that has occurred, e.g. time/temperature changes or production of substances such as ethylene which demonstrates fruit ripening or thermochromatic inks that can change colour within a certain temperature range.

25.5   Whilst many technological applications are in their development infancy, research is developing into the use of sensors, web-services barcodes and the architecture required to access information via phone apps.

25.6   Packaging elements can also be used to prevent theft and removal from storage or retail shelves via electronic article surveillance (EAS) technology. Other technology can be used, such as anti-counterfeiting and anti-tamper devices. These packaging technologies include watermarks, holograms, thermochromatic inks, micro-tags, tear labels and tapes.

26   INTERNAL STORAGE

Principle

Storage areas should be so designed that they are fit for purpose, and the layout and materials used allow for appropriate and effective cleaning. Storage procedures should be in place to prevent damage or deterioration of both premises and the materials contained therein. Effective storage procedures minimise the risk of contamination and are a key prerequisite within a good manufacturing practice (GMP) system. Raw materials, packaging materials, part-processed and finished products, cleaning chemicals, personal items, equipment and machinery spares should be stored, where possible, in segregated, separate storage areas to minimise the risk of cross-contamination.

General

26.1   Access to material and product storage areas should be restricted to personnel working in those areas and other authorised persons. Consideration should be given to site security and entry controls into designated storage areas. Designated separate lockers may need to be provided for both internal and external workwear as well as personal belongings that must not be taken into the production and storage areas (see 17.21f).

26.2   Packaging, food ingredients and products, equipment and other items should all be stored in separate, designated storage areas. Materials and products should be stored under the conditions specified in their respective specifications. Particular attention should be paid to the avoidance of allergenic or microbiological cross-contamination and tainting. Where special conditions are required, they should be regularly checked for compliance with such conditions.

26.3   Materials and products should be stored in such a way that cleaning, the use of pest control materials without risk of contamination, inspection and sampling, retention of delivery identity or batch identity, and effective stock rotation can be easily carried out.

26.4   There should be effective protection of equipment spares, materials and products from contamination during storage.

26.5   Storage areas should be so designed as to minimise the risk of contamination of stored items. Effective cleaning of storage premises and equipment must be carried out at the designated frequency and using the methods and materials specified in documented cleaning schedules and instructions (see Chapter 21). The effectiveness of cleaning should be verified and as a result of such verification any appropriate action taken as necessary to address non-compliance.

Food & Drink – Good Manufacturing Practice: A Guide to its Responsible Management, Seventh Edition.
The Institute of Food Science & Technology Trust Fund.
© 2018 John Wiley & Sons Ltd. Published 2018 by John Wiley & Sons Ltd.

26.6 When developing storage procedures, the following should be taken into consideration:

(a) lighting, temperature and humidity control and ventilation should be adequate for the purpose. Storage areas should be designed and managed to minimise condensation, and any condensate pipes, for example from refrigeration units, should be designed to flow directly above a drain and not be allowed to drip onto the product, materials, packaging, equipment and personnel. This should be verified during premises audits. The condensate pipework should be designed to prevent airflow back into the pipework and consideration should be given to the need to sanitise the condensate produced. The pooling of water around storage areas, especially where there is vehicular access, should be minimised;

(b) storage areas should be kept clean and tidy to minimise harbourage or food sources for pests (see Chapter 22);

(c) damage to stored materials should be minimised and all spillages should be cleared away promptly;

(d) storage areas should have adequate proofing to prevent pest ingress and external doors should not be left open;

(e) all materials within storage areas should be protected from excess heat and light, water penetration and accumulation of foreign matter;

(f) stock rotation should be undertaken, and all items should be marked with their identification so that identification information is clearly visible during storage and traceability of all items is maintained;

(g) items stored on pallets should be neither touching the walls nor blocking the main doors or passageways. If racking is used, the layout should be designed so that the racking is far enough away from the wall to prevent pallets and palletised product being damaged or touching the wall. There should be designated vehicular and pedestrian access in storage areas with racking so that product can be safely inspected during storage. Space should be left in all gangways for product inspection to take place;

(h) any suspect stock should be segregated, ideally in an area designated for that purpose; and

(i) product should not be double stacked where this could prove a potential contamination risk or could affect the integrity of packaging.

Procedures should be developed on the basis of the points raised above, then implemented and verified to ensure that they are effective, understood and followed consistently by those staff working in storage areas. This is especially important in temperature-controlled storage areas such as chilled or frozen stores and where allergens are stored on-site. Furthermore, staff should be trained to understand the need to prevent material, packaging, equipment and product damage during storage and why the maintenance of product safety, integrity and product quality is important. Routine audits, or other appropriate verification activities, must be undertaken and recorded to demonstrate that damage of product and materials during storage is routinely assessed and minimised. Appropriate corrective action should be implemented as necessary in the event of non-compliance and should be followed up to ensure that such actions are appropriate and remain effective.

26.7   The quality control manager or designate is responsible for developing and implementing appropriate monitoring procedures for temperature-controlled storage areas to ensure that the storage area is capable of maintaining the appropriate temperature profile during work activities. The results of monitoring should be formally recorded and any appropriate corrective action taken where required (see Chapter 28).

26.8   Products that have been recalled or returned, and batches that have been rejected for reworking or recovery of materials or disposal, should be so marked and physically segregated, preferably in an entirely separate storage facility (see Chapter 29).

26.9   Material deliveries and product batches temporarily quarantined pending the results of testing should be so marked and suitably segregated, and effective organisational measures implemented to safeguard against unauthorised or accidental use of those materials or despatch of those products.

26.10   If a batch of finished product has to be temporarily stored unlabelled, to be labelled at a later date, the greatest possible care should be exercised in maintaining its exact identity and ensuring correct durability indication when the product is labelled (see Chapter 14).

26.11   Storage areas should be regularly inspected for cleanliness and good housekeeping, and for batches of products that have exceeded their shelf life or, in the case of date-marked products, leave insufficient time for retail display. These inspections should be formally recorded, and the records should include in the event of non-compliance any required corrective action and demonstrate that corrective action has been followed up to verify it has been effective.

26.12   Risk analysis should be undertaken by the quality control manager or designate to ensure that cross-contamination, including where relevant airborne taint, is prevented. This is especially so at times of peak production where storage space may be limited. The Food Standards Agency (FSA) publication *E. coli O157 – Control of cross-contamination: Guidance for food business operators and enforcement authorities* (2014) stresses the importance of managing storage not only of ingredients and finished products, but also of packaging. It also highlights the importance of ensuring that physical separation of materials is effective especially if high-risk, ready-to-eat foods are manufactured/produced.

26.13 Pallet labelling should be undertaken so that it is clearly visible during storage. Attention should be paid to the effective adherence of pallet labels, especially in storage conditions that could affect the ability of the labels to remain intact on the pallet. The number of pallet tickets applied (and on which faces of the pallet) needs to consider the visibility of pallet tickets, especially when product is placed in racking. Consideration should also be given to maintaining traceability on pallets, especially of raw materials, when the outer wrapping is removed and/or where pallet tickets may have been adhered to boxes at the top of the pallet that will be used first. Traceability must be maintained at all times.

Storage of Chemicals, 26.14 The stock controller or designate should be responsible for the
Lubricants and Oils taking into stock of all chemicals. She/he should also be responsible for ensuring that these items are as per the delivery instructions and are held, in storage, in their original packaging until required by particular personnel for cleaning or maintenance activities. All materials must be stored in sealable and labelled containers and handled and transported in a safe and responsible manner. Stores should be sound, secure, bunded (or alternatively the materials can be stored on bunded pallets), well ventilated, frost proof, have ease of access and have sufficient light to enable the operator to read the product label. Appropriate warning signs should be placed on access doors according to the inherent characteristics of the chemicals, for example flammable and corrosive. All materials should be stored off the floor, if not on shelving then on pallets. Shelving should be made from non-absorbent material and powders should be stored on shelves above liquids. When materials are on shelves then the store must be bunded. Stores should be able to retain spillages, and emergency procedures should be in place to deal with accidental spillages. Protective clothing must be worn where applicable. Appropriate personal protective equipment (PPE) should be supplied for all operations involving chemicals, but it should not be stored in the storage area. PPE should be stored in a designated clean, dry, well-ventilated and secure locker. The minimum requirements for PPE are detailed on the chemical product label. Any additional requirements should be identified during the Control of Substances Hazardous to Health (COSHH) assessment. Protective clothing must be personal to the individual, suitable for its intended purpose, in sound condition and the correct fit for the wearer. PPE should be cleaned, maintained, stored and disposed of according to manufacturer's recommendations and statutory requirements. Empty containers must be stored as per legal requirements and must be suitably disposed of.

Product Integrity 26.15 The possibility of sabotage, vandalism, terrorism and other types
and Illicit Behaviour of illicit behaviour (see Chapter 7) means that material, waste and product storage areas may prove vulnerable points for food

crime. Access to material and product storage areas should be restricted to personnel working in those areas and other authorised persons. Consideration should be given to site security and entry controls into designated storage areas. Designated separate lockers may need to be provided for both internal and external workwear as well as personal belongings that must not be taken into the production and storage areas.

27 CRISIS MANAGEMENT, COMPLAINTS AND PRODUCT RECALL

Principle

The full significance of a food safety, integrity or quality complaint may only be appreciated by certain responsible persons, and then possibly only with the knowledge of other related complaints. A procedure must therefore be provided for the appropriate channelling of all complaint reports and the analysis of complaint data. A product defect coming to the manufacturer's attention, whether through a complaint or otherwise, may lead to the need for a product withdrawal from the retail distribution system, or a public product recall also involving return of products by members of the public. There should be a predetermined written plan, clearly understood by all concerned. It should address the withdrawal or recall of a product, or a known batch or batches of product known or suspected to be hazardous or otherwise unfit, or the withdrawal or recall of wholesome but substandard product that the manufacturer wishes to withdraw or recall. A crisis management procedure should be established should such a situation arise and the members of a crisis management team should be defined within the procedure. This procedure is a key prerequisite of good manufacturing practice (GMP).

Complaints

27.1 The system for dealing with complaints should follow documented procedures that indicate the responsible person, and their deputy, through whom the complaints must be channelled.

27.2 If the responsible person is not the quality control manager, the latter should be fully informed and closely consulted. The responsible person should have the appropriate knowledge and experience, and the necessary authority to decide the action to be taken in the event that a complaint arises.

27.3 Where possible, product food safety, legality and quality complaints should be thoroughly investigated by the quality control manager, with the cooperation of all relevant personnel, and a report prepared as a basis for action and for the records. This report could be analysed as part of the management review process (see 11.2). This process should be robust enough to differentiate between one-off incidents and an ongoing trend.

27.4 Corrective action should include responding to the complainant, and if required responding to any enforcement authority involved. Where the complaint is justified, steps to remove or overcome the cause and thus prevent recurrence should be taken, and the defective material that the complaint sample might represent should be dealt with as necessary, including possibly a product withdrawal or recall. Root cause analysis should be undertaken to identify the factors that have caused the incident

Food & Drink – Good Manufacturing Practice: A Guide to its Responsible Management, Seventh Edition.
The Institute of Food Science & Technology Trust Fund.
© 2018 John Wiley & Sons Ltd. Published 2018 by John Wiley & Sons Ltd.

so that appropriate preventive and corrective action is implemented and is effective (see 28.10–28.12).

27.5 Complaint reports should be regularly analysed, summarised and reviewed for any trends or indication of a need for a product recall or of any specific problem requiring attention. It is strongly recommended that appropriate summaries that include comparative data are developed and that they are regularly distributed to directors and senior management and subject to formal management review (see Chapter 11). Trend analysis should relate the number of complaints to volume produced to give a better reflection of performance, such as complaints per 100,000 units or complaints per million units sold/produced. Whether the complaints should be related to products sold or produced will depend on the seasonal variation of production and also the products' indication of durability. Fluctuations in production levels/purchasing levels could, if not taken into consideration, potentially skew the trend analysis for a given month or quarter. Measurable indicators such as complaints per million units sold, percentage compliance with agreed service level or rating of types of complaint should be developed and form part of the management review process.

Product Withdrawal and Recall

27.6 The type of occurrence that constitutes a product incident should be determined. As there are so many types of incident that could occur, it is not possible in this publication to give a definitive list or be specific on the actions to be taken in each instance. The type of product incident likely to occur will be specific to the product and its ingredients, the process employed in preparing, storing and manufacturing the product, and the frequency of reworking or regrading activities, which in turn will reflect ongoing supplier performance at meeting material specifications and the overall effectiveness of the quality management system (QMS), the food integrity management system (FIMS) and the food safety management system (FSMS). An incident could be contamination due to an allergenic, biological, chemical or physical hazard whether accidental or malicious, human contamination such as blood or other material, adulteration and/or any other incident that could affect the quality, safety or legality of the product, such as incorrect labelling or incorrect use of ingredients.

27.7 A responsible person, with appropriate named deputies, should be nominated by senior management to initiate and coordinate all withdrawal and recall activities, to liaise with retailers and to be the point of any contact on withdrawal and recall matters with regulatory authorities as applicable to the incident.

27.8 The design of manufacturing records systems and distribution records systems, and the marking of outer cartons and of individual packs should be such as to facilitate effective withdrawal or recall if necessary. A good system of lot or batch marking will pinpoint the suspect material and help avoid excessive recall. Lot and/or batch marking of food ingredients and products is a requirement of regulatory and market requirements (see Chapter 14).

27.9 There should be written withdrawal and product recall procedures, and they should be capable of being put into operation at short notice, at any time, inside or outside working hours. To ensure this is possible, an out-of-hours contact list should be maintained and include contact details for staff, contractors, suppliers, customers, emergency services, specialist laboratories, legal advisers, certification bodies and government and enforcement regulatory bodies. Procedures should also include a communication plan, which details the communication levels both internal and external to the organisation as well as media communication that would be used in a full product recall.

27.10 The procedures should be shown to be practicable and operable within a reasonable time by carrying out suitable testing of the procedures. The product withdrawal/recall and management of incidents system must be tested at least annually to determine the effectiveness of the system both for traceability from the finished product to raw materials and packaging materials used and tracking from a batch of raw material to the individual batches of products that it was used to make.

27.11 It is important to test the product withdrawal/recall procedure for a range of incidents. For example, a test could be undertaken for glass contamination, metal contamination or an instance of contamination at the supplier and a number of other scenarios. Each type of incident will require a different audit trail, and only by rotating the type of incident that is tested will the organisation be able to demonstrate the true effectiveness of the product recall and withdrawal procedure across a range of potential incidents that could occur. If the product recall test is only undertaken to test the ability to trace a production code to a retail depot or from depot to supplier, it is in effect only a traceability test. Thus, in the event of a glass contamination incident, the ability of the organisation to collate the data and records required will never be tested and could be fundamentally flawed. This will then only become apparent in a 'live' situation. Product recall/withdrawal tests should be documented and records retained. Corrective action identified as a result of non-compliance during a test must be verified to ensure that it has been implemented and has been effective.

27.12 The length of time for a product withdrawal/recall test between initiation and completion should also be recorded to ensure that a potential withdrawal/recall would be carried out in an appropriate timescale.

27.13 The procedures and documentation should be reviewed regularly to check whether there is a need for revision in the light of changes in circumstances, resources or who is deemed to be the responsible person. This could include changes to telephone numbers, contact details or specific customer procedures such as an amendment for the time limit for notification.

27.14 Product withdrawals or recalls may arise in a variety of circumstances, which fall into three main categories:

(a) where the food enforcement authorities become aware of a hazard or suspected hazard, and information and cooperation from the manufacturer or importer is necessitated;

(b) where the manufacturer, importer, distributor, retailer or caterer becomes aware of a hazard or suspected hazard;

(c) where there is no hazard or suspected hazard involved, but there is some circumstance (e.g. substandard quality, mislabelling) that has come to light and that prompts the manufacturer, importer or retailer to decide to withdraw or recall the affected product.

In case (c), the company will itself have to organise the withdrawal or recall operation. In cases (a) and (b), consideration may be given to issuing a public food hazard warning. Generally, this would be done in consultation with the manufacturer or importer, the distributor or retailer, and any relevant enforcement authority interest. The UK FSA issues Allergy Alerts and Food Alerts based on recalls from the retail stage of the supply chain. The FSA issues a Product Withdrawal Information Notice or a Product Recall Information Notice to let the general public, the food supply chain and local authorities know about problems associated with food. A Food Alert for Action is issued where formal intervention by enforcement authorities is required. Normally any arrangements for recall would be discussed so that the most appropriate methods could be effected or endorsed by the relevant authorities, and would also take into account any requirements for or arising from the information indicated below.

27.15 Although a defect or a suspected defect leading to withdrawal or recall may have come to light in respect of a particular batch or batches or a particular period of production, urgent consideration should be given to whether other batches or periods may have also been affected (e.g. through use of a faulty material, or a plant or processing fault) and whether these should also be included in the withdrawal/recall.

27.16 The procedures should lay down precise methods for notifying and implementing a withdrawal or recall from all distribution channels, retailers and goods in transit, that is, wherever the affected product might be. It should also include a procedure to prevent any further distribution of affected goods. The recall procedure should also provide for the method of public notification.

27.17 Notification of withdrawal/recall should include the following information:

(a) name, pack size and adequate description of the product;
(b) identifying marks of the batch(es) concerned;
(c) the nature of the defect;
(d) action required, with an indication of the degree of urgency involved.

27.18   Recalled or withdrawn material should be quarantined pending a decision as to appropriate treatment or disposal.

Emergency Procedure

27.19   Regrettably the possibility of threats arising from the actions of second or third parties must be faced, for example deliberate contamination or poisoning of product or ingredient by terrorists, extremists or otherwise misguided persons (see 5.4, 6.3 and 9.21). Although some of the additional action that might be taken in such circumstances could be considered outside the scope of this Guide, it is included because those concerned in the manufacturing operation would very probably become involved in product incident assessment and, if required, isolation, withdrawal and/or a full recall.

27.20   The first intimation of a problem in this area could come from a whole variety of sources, for example a consumer complaint, the retailer, the media, the police, the enforcement authorities, employees or by telephone, email or texting, post or personal contact with any company location or any employee at any time.

27.21   It is therefore essential that any personnel engaged in manufacture should be aware of company procedures to be followed in dealing with such threats both within and outside of normal working hours, and that suitable arrangements exist for calling in key personnel out of hours in such an emergency. The extent to which any such emergency procedures may override normal lines of management should be explicitly stated. These procedures should be formally documented (see Chapter 6).

27.22   Faced with an emergency situation, the withdrawal/recall procedures described above will apply, while the expertise of those involved in quality control and other relevant functions should be put at the disposal of the crisis management team responsible for handling the emergency.

27.23   The possibility of such sabotage, vandalism, terrorism and even site invasion may indicate a need for particular security precautions in vulnerable areas, for example entrance security, closed-circuit television (CCTV) security, code pads to open external doors to manufacturing areas, locked rooms and the use of tamper-evident or other type of security seals to safeguard raw materials and finished product during storage or transfer (see 9.21).

27.24   Any emergency or recall situation is likely to involve retailers, wholesalers and/or food service, and a smooth and effective interface with their procedures should be achieved as early as possible during the crisis.

27.25 The risk of harm from the actions of both current and previous employees needs to be formally addressed by the manufacturer, including consideration of the need for pre-employment screening and employment termination interviews and protocols.

27.26 An important element of GMP is maintaining the continuity of supply in the event that an incident occurs that disrupts 'normal' manufacturing activity. The business operator should have developed written continuity plan(s) to identify the impact of specific incidents and what actions would be taken in the event that they occurred. Malicious damage and sabotage have been previously addressed in this chapter but the following should also be considered:

- natural disaster, for example flooding, earthquake, hurricane or other weather event;
- environmental release, for example refrigerant, gas or failure in temperature control equipment or as a result of fire on the premises; and
- disruption of major services, including water, energy, transport and logistics, communications, for example Internet and telephone, or staff availability, including a disease or food safety outbreak in the wider population.

27.27 The PAS 96:2017 *Guide to protecting and defending food and drink from deliberate attack* identifies steps that need to be taken to develop food protection and defence procedures and an associated emergency response and business continuity management system (BCMS) to ensure that the business can effectively manage a potential incident (see 5.6). Business continuity plans are an important element of the manufacturer's formal management systems. They could be situated within the QMS, FSMS or the FIMS. BS 11200:2014 *Crisis management. Guidance and good practice* gives guidance to help plan, establish, operate, maintain, monitor, verify and review a organisation's crisis management plan and its wider capability.

ISO 22313: 2014 *Societal Security – Business continuity management systems – Guidance* identifies the requirements for setting up and managing an effective BCMS that aligns with recognised best practice. ISO 22316:2017 *Security and resilience: Organisational resilience – Principles and attributes* provides guidance on resilience, i.e. being able as a manufacturing organisation to anticipate and respond to potential threats and opportunities and adapt to risk and the changing environment in which the business operates. A resilient business can absorb and adapt to supply chain and market shocks and still meet its organisational objectives.

28   CORRECTIVE AND PREVENTIVE ACTION

Principle

A procedure must be provided for appropriate action to be taken in the event of product or service non-conformity that could impact on product safety, legality or other specified quality characteristics. This procedure needs to provide for the corrective actions to be taken on identification of non-conformance, determining the extent and the implications of that non-conformance and the appropriate actions to be taken. Preventive actions are those activities undertaken as a result of identification of a weakness in the management system that has yet to lead to non-conformity, but could if not addressed lead to non-conformity in the future.

General

28.1   Major non-conformance is defined as a food safety management system (FSMS), food integrity management system (FIMS) or quality management system (QMS) failure or fault that could give rise to a food safety hazard, integrity threat or major product quality defect. However, if minor non-conformance persists or if a trend is noted in a series of minor non-conformance, this sequence of events could be deemed major.

28.2   A procedure should be developed and implemented by the quality control manager to ensure that, once identified, non-conforming items are clearly identified, segregated and/or otherwise controlled so that unauthorised or inadvertent use is prevented and disposal is formally agreed and recorded (see Chapters 27 and 29). The procedure should also ensure that all events of non-conformity are fully and adequately recorded.

28.3   Routine inspection, monitoring or verification activities, incorporated within the QMS, FSMS or FIMS, may identify instances of non-conformance. Routine inspection, monitoring or verification procedures must identify the actions to be taken in the instance that non-conformance of product or service is identified.

28.4   Corrective action procedures should be established not only for the manufacturing process, but also during receipt, storage and distribution, including the transport of raw, part-processed and finished products.

28.5   All corrective action should be documented on the appropriate records that identify:

- the date the form is completed;
- details of the actual nature of the non-conformity;
- product details, including traceability information, so that the batch(es) involved can be identified;
- how much material and/or product is affected;

Food & Drink – Good Manufacturing Practice: A Guide to its Responsible Management, Seventh Edition.
The Institute of Food Science & Technology Trust Fund.
© 2018 John Wiley & Sons Ltd. Published 2018 by John Wiley & Sons Ltd.

- location of the material and/or product, for example on-site or through the distribution chain;
- the root cause of the non-conformity (see 28.7);
- the required action to be taken, including the need for product recall or product withdrawal;
- the timescale for the required action;
- who is responsible for undertaking that action;
- that all parties have accepted the non-conformity and agreed the appropriate corrective action;
- who is accountable for the decision on the corrective action necessary; and
- who is responsible for ensuring corrective action has been undertaken and is effective.

28.6 Internal audits may identify evidence of non-conformity and necessitate the requirement for corrective action. If there are a number of corrective actions, the auditor should develop a corrective action plan (see 11.9–11.14).

28.7 Corrective action plans should address the following:

- ensuring all raw materials, ingredients, processing aids, packaging and part-processed and finished products are safe, legal and comply with specifications;
- minimising any impact on the customers and ensuring the non-conformity does not occur again;
- the importance of the non-conformity, i.e. whether the non-conformity is minor, a one-off incident, a repeated minor incident or a major system failure; and
- the resources required to implement corrective action in terms of financial resource, training, personnel and the development of documentation and follow-up activities to ensure it has been effective.

Corrective action plans need to be reviewed by the quality control manager or designate to ensure that the required actions are undertaken by the designated date and the actions are effective. The corrective action plan needs to take into account both the immediate corrective action to ensure that the non-conforming product is brought back under control and the longer-term corrective action that addresses the root cause of the non-conformance. Root cause analysis can take many forms and follow specific mechanisms, but essentially it is a structured management approach that identifies the factors that resulted in the non-conformance in order to determine the most appropriate corrective or preventive action. The factors that could be considered include the actual nature of the non-conformance, the magnitude (major or minor) and the consequences of the problem in order to identify the actions, conditions or behaviours that need to be changed to prevent reoccurrence and/or other similar problems from occurring. Conditions can include processing parameters, procedures, protocols or standards implemented in the manufacturing operation.

28.8 Areas of non-conformity and the status of corrective actions as well as the effectiveness of previous corrective actions provide input to the management review process (see Chapter 11).

28.9 Corrective action must be followed up at an appropriate interval to ensure that the corrective action has been implemented and has been effective. Verification activities to assess effectiveness of corrective action include:

- review of monitoring results, giving feedback on the effectiveness of the corrective action;
- product testing, for example shelf-life testing and microbiological testing; and
- follow-up audits to ensure that the non-conformity identified is now under control.

Customer Complaints/Depot Rejections

28.10 Corrective action arising as a result of customer or consumer complaints (see Chapter 27) and/or customer depot rejections needs to be formally controlled. The quality control manager should develop procedures that define how all quality aspects of distribution activities will be coordinated (see Chapter 32).

28.11 The quality control manager, or designate, should make a formal assessment of the product on its return to determine the correct disposal. The quality control manager should develop an appropriate documented procedure with associated forms to control this process, including the responsibilities of key individuals in the formal decision making. Routine analysis of the records should be undertaken to determine trends and any additional corrective action that may be required.

28.12 If product is reworked or regraded and then redistributed, records must be maintained to ensure traceability to the original batch details and associated records (see Chapter 29).

Preventive Action

28.13 Preventive action arises following identification of a weakness in the FSMS, QMS or FIMS as a result of inspection, monitoring or verification, including audits and management review processes, trend assessment and risk analysis. Preventive action is designed to eliminate the cause of a potential non-conformity and prevent it occurring in the first place. Preventive action is a proactive rather than a reactive activity, and a preventive action plan forms part of the manufacturing organisation's continuous improvement plan. Preventive actions also can be implemented to improve production efficiency, assess new technology or new methods or procedures, improve skill levels and staff knowledge or to reduce repetition in the formal documented systems. Preventive action plans should be established, formally agreed and reviewed at designated intervals to ensure that the preventive actions have been completed and are effective.

28.14 Preventive action is especially important when considering potential criminal activity that could be perpetrated on the manufacturing site (see Chapter 7). Preventing and reducing the risk of criminal activity is more effective than reacting to crime should it occur.

28.15 Corrective and preventive action (CAPA) is a term often used in good manufacturing practice (GMP), hazard analysis critical control point (HACCP) and threat analysis critical control point (TACCP) to describe the processes organisations undertake to eliminate non-conformity and to develop a continuous improvement programme that is appropriate to the products manufactured and the processes employed. Essential within a CAPA process is the use of root cause analysis in terms of potential and actual non-conformity associated with materials, products, processes, people and the formal management systems (see 28.7). The steps of the CAPA process can be described as:

1. Identify the problem, failure or weakness.

2. Determine the root cause.

3. Determine the corrective or preventive action.

4. Implement the action.

5. Review and audit: follow up to ensure action is effective.

6. If the action is ineffective then follow previous steps and implement additional corrective action.

7. Analyse the system for further areas requiring CAPA.

The CAPA process should be fully documented, including maintenance of all records of actions undertaken and their effectiveness. There are multiple tools available to assist in root cause analysis, including Ishikawa or fishbone diagrams, the 5 Whys approach, Pareto analysis and so on.

29   REWORKING PRODUCT

Principle

Material may be recovered, reworked or reprocessed by an appropriate and authorised method provided that the material is safe and suitable for such treatment, and traceability of original raw materials and part-processed product is maintained so that the resulting product complies with the relevant specification and that the related documentation accurately records what has occurred.

General

29.1   Consideration should be given to how reworked material is assessed to ensure that a reworked material is not in itself a potential contaminant. Examples are:

- material from products containing ingredients or additives such as preservatives and colourants not present in the intended recipient product in which the material will be reworked;
- material from products being reworked that do not comply with the specific production method (e.g. halal, kosher, organic, vegan, vegetarian) of the intended recipient product;
- material being reworked not complying with the compositional profile of the intended recipient product (e.g. health products such as diabetic, low carbohydrate or low fat, or products suitable for a low-calorie diet);
- material or intended recipient product of a different specified provenance, country of origin, assurance status or designated geographic status so that identity preservation is lost; and
- material from products containing food allergens being reworked into the intended recipient product where allergens are absent (see Chapter 8).

All reworking activities must be undertaken in line with strict procedures to ensure that the resulting finished product is safe and complies with relevant legal and quality criteria.

29.2   As there are so many different circumstances that can arise with different kinds of food products and processes, it is not possible to be specific here about each of them. The matters referred to here, however, may be classified under three main groups, namely systematic, 'semi-systematic' and 'occasional'. However, as defined in 29.1, in all circumstances appropriate precautions must be taken to avoid microbiological contamination, introduction of undeclared ingredients, cross-contamination with allergens or the introduction of foreign matter, and to avoid the loss of traceability and provenance.

29.3   The possible carrying forward of perishable material left over from the previous day should be subject to formal risk assessment by the quality control manager. Where a quantitatively known

Food & Drink – Good Manufacturing Practice: A Guide to its Responsible Management, Seventh Edition.
The Institute of Food Science & Technology Trust Fund.
© 2018 John Wiley & Sons Ltd. Published 2018 by John Wiley & Sons Ltd.

product residue from the previous production is systematically utilised as one of the starting materials for the same or another product (e.g. dough trimmings in biscuit manufacture), that should be written into the master manufacturing instructions, and the rate or conditions of use there specified should not be departed from other than through the established procedure for varying master manufacturing instructions.

29.4 'Semi-systematic' applies to instances where a variable quantity of intrinsically satisfactory but extrinsically unacceptable product occurs and can be reused (e.g. misshapen or short-weight moulded chocolate bars) or to instances where a usable starting material can be extracted from wholesome but defective product (e.g. recovery of sugar as a syrup from misshapen or erroneously formulated sugar confectionery). In such circumstances, provision for such recovery should be made in the master manufacturing instructions, specifying a maximum limit to the rate of incorporation.

29.5 'Occasional' instances are any instances other than referred to in 29.3 or 29.4. They should in all cases be subject to risk assessment by the quality control manager before any decision as to disposition. The four main categories are as follows:

(a) Batches of intermediate or bulk product that have been quarantined as sub-standard. In some instances (e.g. insufficient salt in a batch of soup or sauce), simple adjustment by addition of the calculated amount of deficient ingredient will suffice; in the case of an excessive amount of an ingredient, rectification may be possible by making a special batch with a calculated deficiency of the ingredient in question and blending the two batches. An alternative in the two cases mentioned, and the only possible use where other types of defect are involved, would be to rework a certain amount of the defective material into a number of succeeding batches. The amount per batch should be determined by experience or a trial where necessary, and agreed by the quality control manager, and should be the subject of a written instruction. Particular regard should be paid to any microbiological or other problems that might arise through holding the material in question for the length of time involved and how traceability can be maintained.

(b) Packed finished product that has been quarantined as substandard. Where the defect is merely soiled, scuffed or badly applied labels, incorrect labelling or stickering, relabelling is appropriate. In instances where the product itself is substandard, it is rare that in-pack reprocessing would be either possible or appropriate (except in the case of certain canned products where safe heat processing has resulted in products the texture of which could be improved by further heat processing). In general, the utilisation of quarantined substandard packed goods (except in circumstances dealt with in (d) below) requires the emptying of the product from the packages. Its utilisation would then be subject to the considerations indicated in (a).

(c) Packed finished product recalled or otherwise returned from distribution. Similar considerations apply to those referred to in (b), with the addition, however, that as the packs concerned have been outside the control of the manufacturer for a period, they should be assessed even more critically and in the light of information as to their age, history and condition.

(d) In the special case of a group of finished product packages that has failed the reference test for filled weight or volume (either on average quantity or on an excessive number of non-standard packages), various ways of rectification are possible. In the case of failure only because average quantity is too low and no package is below the legally acceptable minimum, rectification may be effected by blending the group with another group specially prepared with the average as far above the target as the defective group's average is below it. Where the failure is only through excess non-standard packages, the group may be rectified by sorting to remove all or most of the non-standard packages. Another possibility is relabelling with a lower nominal quantity. Finally, if it is feasible to open and reseal containers (where this can be undertaken without compromising food safety), topping up with material from the same batch may be used to increase average quantity or decrease the number of non-standard packages. The reader is referred to the **Code of Practical Guidance for Packers and Importers; Weights and Measures Act 1979** (HMSO) for detailed guidance in this topic.

Rejection 29.6 Inevitably, rejection of materials described above will be necessary from time to time, and proper means of disposal should be considered and agreed with the quality control manager, the production manager and any other interested parties, such as the purchasing manager or sales/marketing departments. In determining disposal, due regard should be paid to the needs of securing cost recovery, protecting the company or brand name, protecting the public and complying with appropriate legislative or local authority requirements. For example, disposal subject to appropriate safeguards might range from sale to a third party for relabelling or packing as a lower-quality product, to staining and selling for including in animal feed or to seeking local authority condemnation and disposal at their hands. Consideration must be given to compliance with all current European Union (EU) or UK environmental and food and feed safety legislation, including **Regulation (EC) No. 1774/2002 (enacted in the UK by the Animal By-Products Regulations 2003)**. The EU Animal By-Product Regulation 1069/2009 and Regulation 142/2011 came into force in the UK and are implemented by the **Animal By-Product Enforcement Regulations 2013** and the **Animal By-Product (Miscellaneous) Amendments Regulations 2015** as specific to each country within the UK.

Relabelling 29.7 In any relabelling of packs, any identifying marks carried by the original labels should be carried by the new labels, and where the pack carries a durability indication on the label, the new label should carry a date no later than the original durability indication.

30   WASTE MANAGEMENT

Principle

In the Introduction, the intention was made clear to limit this Guide to matters having a direct bearing on the scientific, technological and organisational aspects affecting the quality, legality and safety of products. For this reason, detailed consideration has not been given to the impact of the manufacturing unit and its operations on the external environment. It is, however, acknowledged here that the management of any food manufacturing operation has general responsibility and, in most countries, legal obligations (with which it must be familiar) for these aspects, including the management of waste.

General

30.1   **EU Regulation No. 852/2004** on the hygiene of foodstuffs and **Regulation No. 853/2004** laying down specific hygiene rules for food of animal origin came into force in all Member States on 1 January 2006. They are enforced by the **Food Safety and Hygiene (England) Regulations 2013** in England and by similar legislations in Scotland and Wales as amended. These require that:

(a) food waste, non-edible by-products and other refuse are to be removed from rooms where food is present as quickly as possible to avoid their accumulation;

(b) food waste, non-edible by-products and other refuse are to be deposited in closeable containers, unless food business operators can demonstrate to the competent authority that other types of containers or evacuation systems used are appropriate. These containers are to be of an appropriate construction, kept in sound condition and be easy to clean and, where necessary, disinfect;

(c) adequate provision is to be made for the storage and disposal of food waste, non-edible by-products and other refuse. Refuse stores are to be designed and managed in such a way as to enable them to be kept clean and, where necessary, free of animals and pests; and

(d) all waste is to be eliminated in a hygienic and environmentally friendly way in accordance with the community legislation applicable to that effect, and is not to constitute a direct or indirect source of contamination.

30.2   The treatment/disposal of unwanted by-products (waste materials) must be controlled and comply with all the appropriate environmental legislation in the country concerned (see Chapter 41). In the UK, waste disposal should meet all relevant national and EC legislation. Where necessary, waste should be removed by licensed contractors. If the waste includes trademarked materials, the waste contractor should provide records of material destruction or disposal.

Food & Drink – Good Manufacturing Practice: A Guide to its Responsible Management, Seventh Edition.
The Institute of Food Science & Technology Trust Fund.
© 2018 John Wiley & Sons Ltd. Published 2018 by John Wiley & Sons Ltd.

30.3　　The **EC Directive on Packaging Waste (94/62/EC)** aims to reduce the volume of packaging waste going to landfill sites by setting targets for the recovery and recycling of packaging waste. The UK legislation that implements the EC Directive is the **Producer Responsibility Obligations (Packaging Waste) Regulations 2007 as amended and 2016 amendments**[1] (first came into effect in 1997), which cover aspects of recycling and recovery. Guidance on current requirements and obligations can be found at https://www.gov.uk/guidance/packaging-producer-responsibilities.

The **Packaging (Essential Requirements) Regulations 2015**[2] (first came into effect in 2003) consolidates and revokes all earlier Regulations that relate to the essential requirements for packaging and ensuring packaging material complies with standards set for heavy metal concentration (see 24.11 for more details).

30.4　　Waste management procedures should address the requirements for waste minimisation, reusing the material wherever possible, waste recycling using approved contractors and waste disposal. The frequency of emptying of waste containers should be defined, and verification activities should be undertaken to ensure that the timescales are complied with.

30.5　　Where licensed contractors are required by legislation for the carriage and disposal of waste from the manufacturing site, all appropriate documentation should be maintained.

30.6　　External waste areas should be maintained in an appropriate hygienic state. Internal and external waste containers should be clearly identified, for example with labels or colour coding for different types of waste, and designed and maintained so that they are fully enclosed, are effective in use and afford effective cleaning and, where required, disinfection (see Chapter 21). Consideration should also be given to effective infestation control (see Chapter 22).

Waste Minimisation　30.7　　Waste minimisation audits should identify ways to prevent waste
Audit　　being produced in the manufacturing unit. They should address the following areas as appropriate:

- raw material use and actual versus expected yield;
- level of non-conforming/rejected product;
- level of waste by-products produced per unit of production;
- potable water consumption and the volume of waste water produced per unit of production;
- energy consumption per unit of production;
- packaging usage and volume of packaging waste produced per unit of production;

[1] https://www.legislation.gov.uk/ukdsi/2016/9780111142431.
[2] http://www.legislation.gov.uk/uksi/2015/1640/contents/made.

- chemical usage; and
- sundry and consumable items.

30.8   Waste minimisation audits should be documented with actions identified, responsibility for the action defined and timescales for completion. Verification of the corrective or preventive action required should be completed (see 28.15).

31   FOOD DONATION CONTROLS AND ANIMAL FOOD SUPPLY

Principle

Food manufacturers may, when they have surplus food material or by-products of the manufacturing process, either sell this product as an animal feed ingredient or donate the food to a food donation supply chain. It is critical that these food products when dispatched by the food manufacturer are safe, meet legislative requirements and are of the required quality.

Animal Food Supply

31.1   There have been a number of high-profile incidents in the food supply chain that have been associated with surplus foods, waste foods and by-products from the manufacturing process that via the animal feed route have led to contamination of food products. Examples in the European Union (EU) include the use of meat and bone meal in ruminant animal feed and association with bovine spongiform encephalopathy (BSE), and dioxin via feed then contaminating pork, eggs and poultry products. If a manufacturer is intending to sell material for animal feed they need to ensure it is safe, free from contaminants and meets legislative requirements. A formal hazard analysis critical control point (HACCP) assessment should be undertaken to ensure any potential risks are identified and adequate controls put in place to minimise their occurrence. The controls that are in place should be included in the food safety management system (FSMS) and consideration should also be given to the risk of misrepresentation of the material and risk of food crime (see Chapters 5 to 7).

31.2   Good manufacturing practice (GMP) with regard to second-grade, surplus or 'waste' product is important in the manufacturing environment if this product then goes on to become an input in an alternative food supply chain. Waste should be collected in suitably constructed and identified receptacles, designated for the purpose and labelled as to their contents, for removal from the production area to collection points outside the buildings (see 19.17 and Chapter 30). Routine audits should be undertaken to ensure that training of staff has been effective in making sure they can demonstrate compliance with procedures for handling and storage of food waste or by-products that are going to enter the food or feed supply chain. This audit should have a wide scope and include observations for signs of non-compliance such as not keeping packaging and food waste separate, insufficient traceability of materials that are going to enter the animal feed supply chain and incorrect storage conditions, for example where waste needs to be kept refrigerated to prevent growth and proliferation of micro-organisms that could lead to toxin or mycotoxin formation.

Food & Drink – Good Manufacturing Practice: A Guide to its Responsible Management, Seventh Edition.
The Institute of Food Science & Technology Trust Fund.
© 2018 John Wiley & Sons Ltd. Published 2018 by John Wiley & Sons Ltd.

31.3 Legislation on animal feed is harmonised at EU level. **EU Regulation No. 183/2005** on feed hygiene, as amended by **EU Regulation No. 2017/625**, requires businesses, including manufacturers selling by-products of food production into the feed chain, to be registered or approved.[1] The regulation requires feed businesses to comply with prescribed standards for storage, facilities, personnel and record-keeping. The regulation is enforced through the **Animal Feed (Composition, Marketing and Use) (England) Regulations 2015 and the Animal Feed (Hygiene Sampling etc. and Enforcement) (England) Regulations 2015** with separate but parallel legislation in Scotland, Wales and Northern Ireland. Principal measures with regard to labelling and composition of animal feed that are applicable are:

- **Regulation (EC) No. 1831/2003** on additives for use in animal nutrition;
- **Directive (EC) No. 2002/32** on undesirable substances in animal feed;
- **Directive (EC) No. 2008/38** establishing a list of intended uses of animal feeding stuffs for particular nutritional purposes; and
- **Regulation (EC) No. 767/2009** on the placing on the market and the use of feed.

Legislation on the labelling and composition of animal feed addresses the requirements for information within the supply chain, namely:

- the information to be provided to purchasers on feed labels;
- the nutritional claims that can be made for certain feed products;
- the names and descriptions to be applied to various feed materials (i.e. ingredients either fed singly or included in compound (manufactured) feeds);
- the additives (including vitamins, colourants, flavourings and binders) authorised for use in animal feed;
- the maximum levels of various contaminants (e.g. arsenic, lead, dioxins and certain pesticides); and
- certain substances that must not be used in feed (see the Food Standards Agency (FSA) website for further details).

Donation-based Supply Chains

31.4 As part of an overall waste minimisation strategy, food that is deemed safe and legal, but for some reason has not been dispatched into the food supply chain for which it was originally intended, as a result of excess stock, product close to duration date etc., may be donated to an alternative supply chain. Food rescue, sometimes called food recovery, is the practice of identifying edible food that if intervention does not take place would otherwise enter a waste stream, landfill etc. The food manufacturer may choose to donate or redistribute rescued, recovered or

[1] https://www.food.gov.uk/business-industry/farmingfood/animalfeed/animalfeedlegislation.

surplus food to people in need. Examples of food distribution organisations (FDOs) include social enterprises, food banks, food pantries and community kitchens. These alternative chains are often called donation-based supply chains.

31.5 FoodDrinkEurope have produced food donation guidelines[2] that were endorsed in 2016 by the European Commission's (EC) Standing Committee on Plants, Animals, Food and Feed. The EC issued EU Guidelines on food donation in October 2017.[3] The EU Guidelines define surplus food as 'consisting of finished food products (including fresh meat, fruit and vegetables), partly formulated products or food ingredients, may arise at any stage of the food production and distribution chain for a variety of reasons [such as] foods which do not meet manufacturer and/or customer specifications (e.g. variations in product colour, size, shape, etc.)' as well as production and labelling errors that generate surplus, problems with over-ordering and/or cancelled orders, and issues relating to date marking.

31.6 The EU Guidelines state that surplus food may be redistributed provided that it is fit for human consumption, is compliant with all EU food safety requirements and provides adequate food information to consumers (see Chapter 14). All food donation activities should comply with the general food hygiene requirements of **Regulation (EC) No. 852/2004** on the hygiene of foodstuffs, including registration, and **Regulation (EC) No. 853/2004** when food of animal origin is redistributed, and any food information given should comply with **Regulation (EU) No. 1169/2011** on the provision of food information to consumers, as amended.

31.7 The type of food (dry goods, bakery items, canned, or perishable, e.g. fresh fruit and vegetables, raw or cooked meat and so on) will determine the inherent degree of food safety risk associated with the product and the controls that need to be put in place (see Chapter 3). Traceability of foodstuffs must be maintained through the redistribution process to safeguard the general public so that in the case of a product recall the manufacturer can implement the appropriate procedures (see Chapters 14 and 27). The EU guidance specifies that the following information should be kept as a minimum:

- name and address of supplier, and identification of products supplied;
- name and address of customer, and identification of products delivered;
- date and, where necessary, time of transaction/delivery; and
- volume, where appropriate, or quantity.

[2] http://www.fooddrinkeurope.eu/uploads/publications_documents/6194_FoodDrink_Europe_Every_Meal_Matters_screen.pdf.
[3] http://eur-lex.europa.eu/legal-content/EN/TXT/PDF/?uri=OJ:C:2017:361:FULL&from=EN.

The guidance suggests that the minimum retention period for keeping records should be five years from the date of manufacturing or delivery.

31.8 The manufacturer, as a food donor, needs to consider the capacity of the FDO for storage, transportation up to the point of receipt and temperature control options, allergen management etc. as particular to the products being donated. Logistical issues such as whether the FDO can accept pallets of product, times when staff are at the facility, and access requirements and size of vehicle need to be considered.

31.9 The food manufacturer should ensure that food donation activities are adopted, implemented and verified in line with the management review, internal audit and verification procedures of the organisation (see Chapter 11).

32   WAREHOUSING, TRANSPORT AND DISTRIBUTION

Principle   *Ensuring the safety, legality and quality of food in storage and in the distribution supply chain requires the development of procedures to ensure the preservation of food and minimise the risk of contamination. Verification procedures also need to be in place to ensure that operating procedures are complied with and are effective.*

General

32.1   Effective warehousing operations should be designed to ensure that all products are easily accessible for load assembly as required, to ensure that aisles and assembly areas are planned so that unimpeded movement is possible to and from all parts of the warehouse, to facilitate proper stock rotation (particularly important in relation to short-life and date-marked foods) and to obtain maximum utilisation of available space, consistent with the foregoing requirement. Design of premises should provide separate routes of entry and movement for vehicles and personnel. Designated walkways should be marked in internal and external areas. Vehicles should be loaded and unloaded in a manner that protects the material and/or product being transported. Where the material is susceptible to temperature abuse and/or weather damage, loading and unloading should be undertaken in covered bays.

32.2   Storage and transport of finished products should be under conditions that will prevent contamination, including development of pathogenic or toxigenic microorganisms, will protect against undesirable deterioration of the product and the container, and will assure the delivery of safe, clean and wholesome foods to consumers. This deterioration includes, but is not limited to, contamination from insects, rodents and other vermin, toxic chemicals, pesticides and sources of flavour and odour taint.

32.3   The buildings, grounds, fixtures and equipment of food warehouses and vehicles should be designed, constructed, adapted and maintained to facilitate the operations carried out in them and to prevent damage.

32.4   Pallets should be placed in prescribed places and stored under cover; gangways should be used as such and not as temporary repositories for stocks.

32.5   Product stacking should have regard for all elements of safety. Pallets should be checked periodically for structural integrity, especially incoming and outgoing goods. Cornerboards should be positioned at the corner of each stack, both to make the corner stand out and to protect the product from accidental

Food & Drink – Good Manufacturing Practice: A Guide to its Responsible Management, Seventh Edition.
The Institute of Food Science & Technology Trust Fund.
© 2018 John Wiley & Sons Ltd. Published 2018 by John Wiley & Sons Ltd.

impact damage by high lift and powered pallet trucks. Palletised product should be checked for stability and to ensure there is no product overhang. Shrink wrap may also be used to minimise product movement on the pallet, but care should be taken not to crush the food products through the tension of the shrink wrap. Pallet stacking configuration and pallet labelling should conform to either internal or customer specifications and should be routinely monitored by quality control.

32.6   When assessing the suitability of wooden pallets for use, the following should be considered:

- design and dimensions are fit for purpose;
- maximum weight load;
- signs of damage and the need for repairs;
- use of any preservation materials that could taint the product;
- dryness of the wood;
- missing blocks;
- protruding nails that could cause damage or affect packaging integrity; and
- potential pest contamination of the pallets.

32.7   Pallets of product should be so spaced as to allow proper ventilation.

32.8   A suitable curtain should be provided at all entrances and exits in order to maintain the internal conditions of the warehouse at an appropriate level for the products therein. A risk assessment should be undertaken, which is formally recorded, that identifies whether covered bays for vehicle loading are required. The use of roller doors, sliding doors or strip or air curtains should be assessed and appropriate measures adopted.

32.9   Warehouse and loading dock temperatures, particularly those for chilled or frozen food storage areas, should be kept at an appropriate level to maintain the wholesomeness of the particular foods received and held in such areas. Routine monitoring of temperature should be undertaken to ensure that storage temperatures remain within defined limits. Temperature monitoring records should be maintained. Procedures should be documented and implemented that define the actions to take in the event of a breakdown in the store.

32.10   Lighting should be as high as possible above the product; the smaller the angle of light source from ground level, the smaller is the shadow made by the stack. All glass and hard plastic items should be suitably protected as per the brittle material control procedure (see 19.36–19.42).

32.11   As soon as product damage occurs or is discovered, damaged goods should be placed in a designated area. Consideration should be given as to whether contamination of other products has been possible, for example if stored above other materials in racking when damage occurs. This is particularly important in the event that allergenic materials are being stored or transported. Care must be taken not to expose foods stored in the warehouse to contamination or infestation. Returns from customers must be assessed for contamination or infestation before being placed in a storage area. Returns from customers must be placed in a designated area until a formal review of disposition is undertaken by the quality control manager or designate.

32.12 Only products that have been properly inspected to ensure that the product and packaging are fully acceptable may be repacked into outer packaging. If it is necessary to repack goods of different production codes into the same outer packaging, the package should be marked with an age code that relates to the oldest packet in the case, that is, the shortest product duration.

32.13 Damaged goods that cannot be repacked must be dealt with prior to disposal to prevent their re-entry into the food distribution chain.

32.14 Docks, railway sidings, bays, driveways and so on should be kept free from accumulation of debris and spillage.

32.15 Fire exits should be checked on a routine basis to ensure that they are kept clear and allow access. Fire appliances should be suitable for use on the commodities concerned and a sufficient proportion of them should be capable of dealing with electrical and petroleum/fuel oil fires. Fire appliances should be checked by a suitably qualified individual or contractor at least annually. Certificates of inspection should be maintained.

32.16 Forklift and other trucks used within the warehouse should normally be battery driven or otherwise equipped to prevent fume or fuel contamination. Procedures should be in place to undertake regular inspections of forklift trucks to minimise the potential for product contamination and records should be retained of the inspections undertaken. Only competent personnel with suitable training should drive the forklift trucks.

32.17 Vehicles, particularly those used for the transport of chilled or frozen foods, should be capable of achieving, and be operated at, temperatures appropriate to maintain the wholesomeness of the foods being carried therein. Routine monitoring of temperature should be undertaken to ensure that vehicle temperatures remain within defined limits. Temperature datalogging equipment should be used where deemed necessary to demonstrate that the vehicle temperature has remained within specified parameters during distribution. Temperature-monitoring records should be maintained. Vehicle procedures should document and implement:

(a) maintenance and hygiene procedures, including checks of vehicles to identify any potential source of odours that could taint product. The requirements for designated loads and the restrictions on mixing of loads should be defined;

(b) actions to be taken to ensure security of the load and in the event of a mechanical or temperature breakdown; and

(c) records that need to be maintained.

When developing hygiene procedures for vehicles, the cleaning of hoses and coupling points must be addressed and hoses should be capped when not in use. Care must be taken to ensure that during loading/unloading the cap is not in contact with the ground or otherwise contaminated. With bulk deliveries, and where required by the manufacturer or customer, records should be retained of the last three loads that have been carried on the trailer/tanker and the cleaning that has been undertaken. The haulage company must comply with any haulage exclusion policies.

32.18   All vehicles, containers and so on should be free from rodents, birds and insects or contamination from them, free from odours, nails, splinters, oil and grease, and accumulations of dirt and debris, and should be in good repair, without holes, cracks or crevices that could provide entrances or harbourages for pests.

32.19   Prior to loading, the vehicle interior (including walls, floor, ceiling and light(s) if internal lights are present) should be inspected for general cleanliness, freedom from moisture, foreign materials and so on, which could cause product contamination or damage to the packages. Lights should be checked to ensure they are intact. The inspection should be formally recorded and linked to the registration number of the vehicle to ensure traceability of information as well as the delivery note (where the vehicle number should also be recorded).

32.20   The load of goods must be evenly distributed so as not to cause the gross weight, or any one of the axle weights of the vehicles, to be exceeded. Where a vehicle is loaded with multiple drops for distribution centres, a load plan should be completed that identifies the position on the vehicle of each pallet. One copy should travel with the vehicle and one copy should be retained for the quality records.

32.21   The load should ride satisfactorily and safely when on the move in order to avoid damage to packages, people or the vehicle in the event of violent braking or cornering.

32.22   Vehicles bringing product to a warehouse should be inspected before offloading for evidence of damage, or of insect or rodent infestations, objectionable odours or other form of contamination. The checks should be documented and include making sure that any lighting or other brittle item is intact.

32.23 If damaged product is accepted and offloaded from the vehicle, it must be kept separate from other product and handled in a manner that will not expose other foods on the vehicle, or subsequently the food warehouse or storage area, to contamination or infestation.

32.24 A procedure should be set up to deal with consequences of accidents and damage occurring when goods are in storage or distribution, for example spillage procedures, salvage or condemnation following damage to goods in a road traffic accident.

32.25 Security precautions should include means of preventing and deterring any tampering with goods in storage and distribution such as the use of tamper-evident seals (see 7.3). Tools, such as digital photographs, are often used to demonstrate the level of security on release of materials by the supplier before entering the distribution chain. These photographs are then emailed to the manufacturer prior to arrival. The delivery can then be checked on arrival at the manufacturer for integrity and signs of tampering.

32.26 Where quality activities are being undertaken during food distribution, such as fruit ripening or meat maturation, the quality control manager must develop documented procedures and associated forms to ensure that the process is suitably controlled and that an effective inspection and sampling plan has been developed for both in-transit activities and inspection on arrival at the manufacturing site.

32.27 Where warehousing is contracted out, the premises, vehicles and conditions should be subject to the manufacturer's food control checks and assessed under the supplier assessment and performance monitoring procedures.

32.28 Where contracted transport is used, documented procedures and/or terms and conditions should be implemented and performance should be assessed under the supplier assessment and performance monitoring procedures.

32.29 In the event of an outbreak, or a suspected outbreak, of a notifiable disease such as foot-and-mouth disease, avian influenza, Newcastle disease or swine fever, current legislation must be complied with throughout the distribution chain, including compliance with movement restrictions in surveillance zones.

32.30 The BRC Global Standard for Storage and Distribution, introduced as a private standard in 2006, has been developed to reflect best practice and ensure product integrity is maintained. Elements of the standard include senior management commitment and continuous improvement, hazard and risk analysis, personnel standards, quality management system, site and buildings standards, vehicle operating standards, good operating practices and facility management.

33   CONTRACT MANUFACTURE AND OUTSOURCED PROCESSING AND PACKAGING

Principle

Where complete or part manufacture is carried out as an own-label, private-label, distributor's-own-brand, contract packing or similar operation, the obligation is on the contract acceptor (the actual manufacturer) to ensure that outsourced processing and production is carried out in accordance with good manufacturing practice (GMP) in the same way that would be expected where she/he manufactures for distribution and sale on his/her own account, except where responsibility is specifically excluded by mutual agreement between the contract giver and the contract acceptor.

Contract Acceptance   33.1   The contract acceptor should ensure that the terms of the contract are clearly stated in writing and that raw materials, in-process materials and end products are covered by adequate, comprehensive specifications (as outlined in other chapters). Any special GMP requirements should be clearly emphasised, and quality control, record transfer, coding, rejection, dispute and complaint procedures should be identified and agreed. Items of possible confidentiality should be identified and any appropriate safeguards be mutually agreed.

33.2   If the contract giver has a designated approved suppliers list, this must be discussed and agreed with the contract acceptor. The responsibility for approval of packaging and artwork design with the packaging supplier must also be agreed. Consideration must be given to the security of packaging and artwork design to prevent inappropriate use and the risk of counterfeiting or other forms of food crime (see Chapters 5 to 7).

33.3   It is normal practice for contract givers to impose contractual conditions that ensure compliance with food safety, legal and quality standards and principles of GMP. This is desirably achieved, at least in the first instance, by a visit to the manufacturing unit by the contract giver's quality control manager and/or brand owner/retailer product technologist as appropriate to the status of the contract giver. The visit should include the following objectives:

(a) to ensure that within the manufacturing environment the food can be produced safely and legally, and to the required quality standards;

(b) to agree a detailed product specification covering all aspects of the product, processing and packing requirements as well as controls for delivery, embracing parameters to be used for acceptance or rejection, and any legal requirements relating thereto;

(c) to agree levels of sampling of raw materials, in-process materials and finished products by the processor and sampling plans to be used in case of dispute;

Food & Drink – Good Manufacturing Practice: A Guide to its Responsible Management, Seventh Edition.
The Institute of Food Science & Technology Trust Fund.
© 2018 John Wiley & Sons Ltd. Published 2018 by John Wiley & Sons Ltd.

(d) to evaluate the adequacy of the control resources, systems and methods, including traceability and records of the manufacturer; and

(e) to agree, wherever possible, objective methods of examination, while subjective measurements should conform to recognised and accepted standards if possible.

Agreement in all five areas is essential for any manufacturer/customer trading relationship and should benefit both parties.

33.4   In the event that the contract giver is outsourcing part of the manufacturing of a branded product, then the brand owner should be notified and approval sought where required. The contract giver should also establish validation, monitoring and verification activities both prior to and for the duration of the contract with the contract acceptor. The extent of these activities will be based on a formal, documented risk assessment and may include, but is not limited to, organoleptic, chemical and microbiological assessment. The contract acceptor will be approved and monitored in line with the contract giver's supplier approval and performance monitoring procedure (see Chapter 23).

Third-party   33.5   Historically, contract givers undertook the verification of manu-
Certification   facturing units against their individual company-specific standards, protocols and codes of practice. Indeed this is still the practice for some contract givers who have developed their own bespoke manufacturing standards. However, increasingly verification is undertaken by third-party certification bodies, especially as the supply chain becomes more global. The development of private standards such as British Retail Consortium (BRC) Global Standard for Food Safety, BS EN ISO 22000 2005, and latterly the work of the Global Food Safety Initiative (GFSI), has seen a rise in third-party certification. This certification is often a prerequisite to supply. It should be noted that acceptance of third-party certification, or the use of third-party auditors, as the sole evidence of suitability to pack or manufacture food products renders the contract giver potentially unable to provide an adequate due diligence defence. There is a requirement for the contract giver to verify that third-party certification and/or third-party inspections and audits are undertaken to the required standard and that the decisions made by such auditors are appropriate and reflect actual practice. In this context, contract givers must determine the level of verification risk (VR) of accepting third-party certification and auditing/inspection. As VR increases, so ultimately does the risk of customer complaint, product withdrawal and recall and incidence of food-borne illness and food poisoning.

33.6   The evaluation for these aforementioned private standards must be undertaken by certification bodies who are accredited against the standard *ISO/IEC 17021-1: 2015 – Conformity assessment – Requirements for bodies providing audit and certification of management systems.*

34   CALIBRATION

In order to ensure the validity of measuring inspection and testing equipment (MITE) that is being used to verify firstly that the product is within specification, safe, authentic and meets current legislation and secondly that processing equipment is maintaining or monitoring process conditions, it should be routinely calibrated to a recognised national standard. During the validation process, the location and type of MITE and the required degree of accuracy should be determined, documented and then monitored and verified to ensure that process and product parameters are continuously maintained within designated boundaries. This is especially important at process steps that are identified as critical control points (CCPs) within the hazard analysis critical control point (HACCP) plan or critical quality points (CQPs) in the quality plan. Equipment calibration is therefore a key prerequisite within a food safety management system (FSMS), food integrity management system (FIMS) and the quality management system (QMS) that collectively incorporate the requirements of good manufacturing practice (GMP).

General

34.1   Calibration procedures should ensure that all equipment required for measuring, inspection, testing and process control is suitable to demonstrate conformance to specified requirements, is regularly calibrated as necessary and its use(s) specified. Calibration procedures should also ensure that there will be no reduction in product quality in the event of failure of any item of inspection or test equipment. The quality control manager should be responsible for the effective design and implementation of equipment calibration procedures. These procedures should identify the actions to be taken in the event that calibration activities indicate that unsafe, illegal or out-of-specification product may have been manufactured and distributed. All calibration activities should be traceable to recognised national standards. The need for equipment sanitisation following calibration activities should be considered and appropriate procedures put in place to minimise the risk of product contamination.

34.2   Equipment may include, but is not limited to:

- weighing scales, weigh cells or checkweighers;
- standard test weights;
- counting devices, for example magic eyes;
- flow meters;
- measuring equipment, for example callipers and sizing rings;
- temperature probes;
- metal detection equipment;
- penetrometers;
- refractometers; and
- pH meters.

Food & Drink – Good Manufacturing Practice: A Guide to its Responsible Management, Seventh Edition.
The Institute of Food Science & Technology Trust Fund.
© 2018 John Wiley & Sons Ltd. Published 2018 by John Wiley & Sons Ltd.

34.3 Calibration may be either an internal activity undertaken by trained personnel or an external third-party calibration service. A MITE register or equivalent document should be maintained for all items of inspection and test equipment other than that which is 'for indication only'. The MITE register should contain the following information as applicable:

- equipment type;
- serial number;
- internal company number;
- equipment location; and
- date last calibrated and date next calibration is due.

For each piece of equipment that requires calibration, a calibration record should be maintained detailing as applicable:

- details and description of the equipment;
- serial number or identification number;
- acceptable degree of accuracy required;
- the range in which the piece of equipment should be calibrated. The calibration range must reflect the range of temperatures, weights and so on within which the MITE will be required to perform during routine monitoring and verification activity;
- the location of the equipment;
- the procedure for calibration (internal or external);
- frequency of calibration;
- the next due date for calibration; and
- evidence of calibration – a valid certificate that is traceable to national standards.

The frequency of calibration and the degree of accuracy required when calibration is undertaken should be determined by risk assessment that is formally documented and routinely reviewed to ensure that product safety, quality and legality can still be maintained. Work instructions and/or task procedures should be developed that outline the protocol for internal calibration activities. Only trained competent staff may undertake internal calibration activities, and training records should be available that demonstrate the level of training of staff. Calibration records should be routinely verified by the quality control manager or designate to ensure that:

- all calibration is up to date;
- all equipment is working to the required level of accuracy; and
- equipment currently in use has a valid calibration certificate and is traceable to a recognised national standard.

A calibration diary is a useful tool to ensure equipment is calibrated on time, especially where there is both internal and external calibration undertaken.

34.4   Once a piece of equipment has been calibrated either internally or externally, it should only be adjusted by authorised personnel and according to prescribed procedures. Any adjustments should be formally recorded. The equipment should bear a formal identification mark, stamp or serial number that is traceable to a calibration certificate and the aforementioned MITE register. All internal weights that are used for daily calibration of weighing scales and checkweighers should be individually marked and traceable to calibration certificates and the MITE register.

34.5   Internal calibration and any required actions should be formally recorded. All MITE should be suitably protected from any damage, deterioration or misuse during handling, maintenance or storage.

34.6   All items that require calibration for which current calibration certificates or their equivalent are not available should be clearly marked as such, for example 'Not calibrated, not in use'. Any equipment marked as such should not to be used for any inspection or testing purposes.

34.7   Equipment that is calibrated independently for which there is no back-up should be calibrated when the company is not in operation (i.e. weekends, holidays).

34.8   Equipment failing calibration should be repaired or replaced. Repaired items should then be recalibrated before reuse. In the event of any item failing calibration, the quality control manager should assess the validity of previous inspections.

34.9   The quality control manager should ensure that there is back-up equipment or an alternative procedure ready for operation in the event of any failure of inspection and test equipment. Back-up equipment should also be incorporated into the MITE register and maintained in working condition.

34.10   The quality control manager should develop procedures to ensure that in the event that inspection and testing equipment is found to be functioning incorrectly or damaged, all product produced since the last satisfactory check can be identified, isolated and retested. This will be a factor in determining the frequency of calibration of items of equipment. The need for product withdrawal/recall should be assessed where appropriate, and action taken to control any affected product (see Chapter 27).

34.11   All scales and checkweighers should also be calibrated and serviced routinely by an external contractor. The contractor should be monitored as per the supplier approval and performance monitoring procedure. Where automatic checkweighers are used, it should be verified at the start of production that they will reject low-weight packs and then monitored at routine intervals during the production shift. If the automatic checkweigher fails to reject the low-weight samples, the quality control manager should be informed and appropriate action taken. It is important that the low-weight samples are identified in such a way that they cannot be confused with actual production packs.

34.12 If there is a failure during routine monitoring of the check-weigher, a back-up manual system of checkweighing must be developed so that it can be introduced until the equipment can be serviced and checked by an appropriately qualified person. All production packed since the last satisfactory check should be identified and rechecked manually to ensure its conformity with specifications. Any out-of-specification product must be suitably controlled.

Principle

Ensuring the safety, legality and quality of products requires the development of testing and inspection procedures to ensure that the food products are safe, wholesome, comply with relevant legislation and meet consistently designated specifications. Testing and inspection procedures should enable the relevant parameters to be monitored so that corrective action can be taken if results fall outside specified limits. The term 'quality control staff' is used below, but it should be recognised that individuals with alternative job titles may undertake these tasks within the manufacturing environment depending on the specific allocation of tasks. This chapter should be read bearing this caveat in mind.

General

35.1   Quality control staff need to be aware of the specified target levels and tolerances, critical limits and control limits for the raw material, ingredient or product being tested. Good laboratory practice is addressed in Chapter 38. Any external laboratory services used must be monitored as per the supplier approval and performance monitoring procedure (see Chapter 23).

35.2   Quality control personnel should have the authority to hold product considered to be outside specified parameters or limits. They must then refer the matter to the appropriate manager for his/her formal decision on disposal.

35.3   Product control, testing and inspection checks should be scheduled throughout the manufacturing process, including intake, on-line and despatch checks. Quality control and/or process records must be completed for all batches/consignments at the time the test, check or inspection is undertaken and should be held on file for both internal reference and customer or third-party auditor review. Historical control and process records should be held according to the controlled records list (see 13.10). Any changes to these records must be authorised by an appropriate individual and the reasons for the change recorded as well as the name of the individual who approved the change.

35.4   Quality control procedures should be developed by the quality control manager considering the following:

- the product characteristics to be assessed and the acceptance criteria;
- the 'volume' of product that constitutes a batch and the information that can be used to identify that batch and where it should be recorded;
- the sampling accuracy and whether the sample is indicative of the whole batch;

Food & Drink – Good Manufacturing Practice: A Guide to its Responsible Management, Seventh Edition.
The Institute of Food Science & Technology Trust Fund.
© 2018 John Wiley & Sons Ltd. Published 2018 by John Wiley & Sons Ltd.

- the equipment required, the appropriate measurements to be taken and the degree of accuracy required for the equipment;
- the skills, training and qualifications required by quality control personnel;
- the points in the process where the measurements will be undertaken and the frequency of checks;
- the actions to be taken in the event of non-conformance;
- the documentation to be completed that records the results of inspection; and
- the requirements for verification of the results of product testing.

35.5   Product characteristics/criteria to be assessed at each process stage should be defined in a quality plan, food safety plan or equivalent. Product characteristics/criteria may also be described in individual customer specifications and could include:

- sample size, which may be determined by legislation, customer requirements or internal procedures. It is important to ensure that the sampling is large enough to be representative of the batch being tested;
- product temperature, especially where it is critical to control food safety;
- traceability as required by the specification;
- provenance: origin, livestock breed and variety. If there are varietal differences such as with produce, especially where there is a legal requirement to label the product with varietal declaration, this must be adequately controlled. The material must be checked to ensure that it is from an approved source, is of the correct provenance if this has been determined as a requirement and is as specified in accompanying documentation. If the material is specified as being from a certified source, for example organic, kosher or farm assured, then identity preservation must be maintained through the manufacturing process;
- quality and appearance. The product must be assessed as per internal/customer specification, including, as appropriate, intrinsic standards such as smell, colour, size and shape, maturity, eating quality, freshness, firmness and cleanliness or level of soiling;
- specific product characteristics that are applicable to individual ingredients and products, for example pH, water activity (a_w), sugar content (% Brix), salt content and preservative levels; and
- packaging in terms of compliance with specification, signs of damage, correct labelling and product information (see 35.11).

35.6   Any material or product failing to meet criteria defined within the specification must be placed on hold and suitably marked and/or held in a quarantine area awaiting a decision on its disposition.

35.7   If the material or product fails to meet the specification at any stage in the process, then it should be held and the quality control manager informed. (S)he should decide, with other members of the management team as necessary, the final product disposal. Disposal will be to reject, to use under concession, to supply to an alternative customer whose specifications it meets or to reprocess or rework material. The appropriate action should be taken only on the written authority of the quality control manager. Causes of non-conformance should be recorded and analysed, and corrective action initiated to prevent recurrence. Corrective action needs to be documented and followed up to ensure that it has been completed and is effective (see Chapter 28).

Quantity Control   35.8   Quantity control systems can be based on average quantity or minimum quantity. The procedure for quantity control should meet the minimum requirements of legislation and also the customer requirements as defined in the product specification. Quality control and production personnel must be aware of the minimum gross pack weight/volume required for each individual pack/production line (including stated product weight/volume, packaging tare weight and any weight loss allowance for product that could dehydrate over its shelf-life duration). Packed product may also be sold to a specific count. Where bulk product is not subject to legislative requirements, it must conform to the relevant specification.

35.9   All weighing scales and weighing equipment should be tested daily against standard test weights that have themselves been calibrated and are traceable to national standards. The results should be recorded on the appropriate record. Before commencing work, the production and/or quality control personnel, as applicable, should ensure that the equipment they are using is working correctly and is suitable for use. At the end of the production shift, they should ensure that the equipment is left in good working order. In the event that equipment is broken, the quality control manager, or designate, should be informed so that (s)he can arrange for the equipment to be repaired and determine the need for product to be held, withdrawn or recalled.

35.10   The use of statistical process control (SPC), especially with automatic filling machines or checkweighers, can assist in the minimisation of 'giveaway' and ensure effective control of product weights and thus productivity. The roles and responsibilities for quality control and production personnel in implementing the quantity control systems need to be formally defined and routinely assessed to ensure compliance with legislative requirements and product specifications.

Label Control   35.11   Product labels should be kept in a designated area until required by production. Procedures should be in place to prevent unauthorised use of labels and the risk of fraudulent activity associated with a given product. Labels should be issued from this area to authorised personnel, and any unused/returned labels should be signed back in and accounted for at the end of production. During product changeovers, all packaging and labelled materials, especially flash or promotional labels, should be removed from the line and the line should be approved by the line supervisor or designate that it is ready to start manufacture of the next product. In the event that labels from previous production runs are found on the production line, the quality control manager should be informed immediately and appropriate action taken, including the potential withdrawal of product from the distribution system. All actions taken should be documented.

35.12 Quality control personnel should approve labels prior to use. The label should be checked by at least two and preferably three competent individuals before it is approved for production. This should be recorded on the appropriate quality record. The quality checks should include as applicable to the label and relevant product specification:

- label format (portrait or landscape) and position on the pack, especially in relation to additional flash or promotional labels;
- product label design and descriptions, including brand owner's name and address, quantitative ingredient declaration, nutrition labelling and claims;
- print registration and quality;
- colour coding (where applicable, e.g. on box end or tray labels);
- product, variety and country of origin;
- volume, weight, size or count;
- bar code (readability to be confirmed by use of an approved bar code scanner and verifier);
- customer department/commodity number (where applicable);
- processing plant/supplier identification number/code;
- durability date: use by, display until or best before, or a combination of the latter two, depending on specific customer requirements;
- price (where applicable); and
- print clarity through the production run and correct size of print.

The first and the last label of product, pack and tray end should be checked from each batch (reel) of label, signed off and affixed to the quality control record. If film is printed on-line, then samples of printed film should be retained. Labels should be checked at designated intervals during the production run for continued conformance. The frequency of checks could relate to a period of production time or production volume, for example every 30 minutes, every hour or every pallet.

Shelf-life Assessment 35.13 Samples should be held for the designated time period and tests carried out as detailed in individual internal/customer specifications. If any rapid deterioration of the product occurs, this should be referred to the quality control manager or designate so that appropriate action can be taken. Shelf-life samples may be held at different temperatures, including frozen, chilled, ambient or accelerated storage. Shelf-life records should be retained and analysed for potential trends. In the event of failure of a given batch to meet safety or quality criteria during the duration coding of the product, appropriate corrective action should be implemented.

Despatch Quality Control Procedures

35.14 Quality checks and the temperature of all products must be taken prior to despatch and noted on the appropriate record. If the finished product is subject to positive release checks, then these must all be undertaken, recorded and signed off if fully compliant so that the batch/load can be formally released for distribution. Positive release procedures should be documented and implemented, and only authorised personnel may formally release a product as specified in the procedure(s). The traceability information on these records should concur with those on the delivery and despatch notes.

Food Safety Monitoring and Verification

35.15 Food safety monitoring and verification activities will be undertaken in line with the prescribed requirements of the food safety management system (FSMS) and/or hazard analysis critical control point (HACCP) plan. The testing undertaken at both control points and critical control points (CCPs) should demonstrate that the process or product complies with the required critical limits and/or target levels and tolerances where these have been defined. Verification activities will also include product inspection and testing activities such as shelf-life testing, testing of products from the retail shelf and other tests as identified by the FSMS.

Principle

The term 'food provenance', as outlined in Chapter 15, relates to not only the geographic elements of where the ingredients and the final food are grown processed and finally manufactured but also how that food is produced and whether the methods of production comply with certain standards and protocols. Authentic products are those that demonstrate a given connection to a recipe, location or social characteristic. Product integrity refers to the wholeness, completeness or soundness of a food product. This chapter considers the emerging testing procedures that are being used to test and verify either product integrity or provenance characteristics that are attributed to the product.

General

36.1   Authenticity is the innate quality of being authentic, genuine and of undisputed origin. Concerns with regard to food authenticity and the potential for substitution and adulteration in the food supply chain mean that increasingly food manufacturers are developing verification activities to determine the provenance and integrity of food. These tests are expensive and need to be completed using sophisticated and expensive equipment, but purchasing this equipment is often outside of the financial resources of many food manufacturers. The quality manager or designate, on behalf of the food manufacturer, needs to undertake a risk assessment to determine which foods are most at risk of food fraud (see Chapter 6) using approaches such as threat analysis critical control point (TACCP). The contract analysis laboratory, if used, must comply with established principles of good control laboratory practice (see Chapter 38). Verification schedules need to be developed, implemented and reviewed to ensure that they are effective. In the event of identification of non-compliance, appropriate corrective action procedures need to be implemented and if necessary a product withdrawal or recall instigated.

36.2   Food authenticity testing uses a range of techniques, including gas chromatography (GC), ultraviolet (UV) spectroscopy, high-performance liquid chromatography (HPLC), liquid chromatography, nuclear magnetic resonance (NMR), mass spectrometry (MS), deoxyribonucleic acid (DNA) fingerprinting and infrared spectrometry techniques such as near infrared (NIR) or mid infrared (MIR). Physiochemical techniques include fluorescence spectroscopy, nuclear magnetic resonance coupled with mass spectrometry of isotopic ratio (NMR/MSIR), ion exchange chromatography/atomic absorption spectrometry (AAS), site-specific natural isotope fractionation by nuclear magnetic resonance (SNIF-NMR), mid- and near-infrared spectroscopy (MIRS–NIRS), Fourier transform mid-infrared spectroscopy

Food & Drink – Good Manufacturing Practice: A Guide to its Responsible Management, Seventh Edition.
The Institute of Food Science & Technology Trust Fund.
© 2018 John Wiley & Sons Ltd. Published 2018 by John Wiley & Sons Ltd.

(FT-MIRS), Curie point pyrolysis coupled to mass spectrometry (Cp–PyMS) and electronic NOSE coupled with mass spectrometry.

Genetic Fingerprinting 36.3 Genetic fingerprinting can provide a unique DNA profile for a given food by identifying selected genetic markers. DNA profiling requires methods such as polymerase chain reaction (PCR) to identify a given species, e.g. with fish or meat. PCR has also been used to determine the livestock origin of cheese and types of cereal. Identification of variety, breed or cultivar can be determined through the use of single nucleotide polymorphism (SNP) or PCR-length polymorphism (PCR-LP). Methods such as SNP and short tandem repeats (STR) can be used to verify breed and species.

Geographic Origin 36.4 PCR has been used to authenticate designated origin products such as those associated with Protected Geographic Indication (PGI) or Protected Designation of Origin (PDO). NIR has been used to determine geographic origin of fish, fish speciation, adulteration of honey with sugar and whether water has been injected into meat, and to differentiate between grape varieties, geographic origin and designation of wines. Chemical composition analysis through the use of isotope or trace element analysis identifies a 'fingerprint' that is specific to a given location. High-resolution mass spectroscopy (HR-MS) determines the isotope ratio, which can be used as a marker for the food. Comparison of isotope ratios can be undertaken using isotope ratio mass spectrometry (IRMS) or by determining the correlation of single or multiple elements in a food, sometimes called multi-element profiling. Isotope profiles and the associated isotope mapping of supply chains has developed rapidly as a tool to verify provenance, product integrity and the authenticity of materials and products. NMR has been used to identify the country of origin or region of production of a range of products, including coffee.

37   LABELLING

Principle

The labelling and presentation of the product, which includes any information printed on or implied by the shape or design of the packaging or by the application of labels to the packs, must correctly correspond to the product concerned, and must comply with the legislative requirements of the country in which the product is to be sold.

General

37.1   In many manufacturing units there will be some combination of different products, different sizes of a given product and different versions of a given product (e.g. variations in composition, branding or language), and it is essential to ensure that the correct supply of packaging and/or label is selected. The correctness of selection should be independently checked by the production management team and quality control before a production run, and any residual material from a previous production run removed from the production area (see 35.12). This should be controlled by formal documented procedures that are verified on a regular basis.

37.2   When a new pack or label design is introduced for a product, residual stocks of the obsolete packaging or label should be destroyed at the designated changeover time. Similarly, where packaging or labels are reference coded or date marked before use, surplus material left from earlier production and bearing a reference or date that is no longer valid should be withdrawn from the production area and destroyed. Redundant packaging can incur significant costs and needs to be effectively managed.

37.3   A particular source of selection error may be the circumstance of material being packed and/or processed in unmarked containers, which are then stored unlabelled for a period of time before being labelled or placed in further containers carrying the required information. All such stored material should be identified with full information to enable it to be correctly labelled in due course, with adequate precautions to ensure that its full identity is not lost (see Chapter 14).

37.4   The task of designing packaging or labels that while fulfilling the presentation wishes of the manufacturer also comply with the requirements of the law of the country for which the product is intended is a complex and difficult one. Legislative requirements governing the labelling of a product are rarely to be found in a single law. In addition to general labelling regulations, the compositional regulations for some products contain some special labelling requirements, which sometimes replace and sometimes add to the provisions in the general labelling regulations. Some

Food & Drink – Good Manufacturing Practice: A Guide to its Responsible Management, Seventh Edition.
The Institute of Food Science & Technology Trust Fund.
© 2018 John Wiley & Sons Ltd. Published 2018 by John Wiley & Sons Ltd.

aspects of labelling requirements are also governed by weights and measures legislation, and various consumer protection laws, so the manufacturer needs to be familiar with these. In addition to regulations, account needs to be taken of case law on the way in which regulations have been interpreted by courts, guidance from official sources and relevant industry codes of practice.

37.5 For a particular product, account has to be taken of all the different requirements affecting its labelling and packaging, not only in terms of the information included but also in terms of its spatial layout, presentation and design. In many cases, interpretations have to be made of how the relevant provisions apply in the particular combination of circumstances presented by the product. Additionally, changes are continually taking place in labelling legislation and there are usually transitional periods when the old and the new regulations operate side by side, requiring detailed knowledge of the changes and the precise timetable. It must be recognised that ensuring compliance is not a task that can be left to graphic designers, advertising agencies, marketing managers or product managers, but requires a considerable degree of up-to-the-minute technico-legal expertise. Such expertise is an essential part of food control and is within the role of the quality control manager with such assistance as she/he may require.

37.6 When a new label for a product is being designed, or an existing one is being revised, whether expert help is sought from inside or outside the company or both, that label must comply with the legal requirements of the country in which the product is to be sold. Final adoption of the label should not be made without the approval of the quality control manager.

37.7 In the UK and the other European Union (EU) Member States, most food labels are legally required to carry a durability indication giving 'the date up to which the food can reasonably be expected to retain its specific properties if properly stored' plus a statement of 'any storage conditions which need to be observed if the food is to retain its specific properties until that date'. This necessitates the manufacturer establishing for each product a shelf life, that is, the duration from the date of manufacture that determines the date to be given on the pack.

37.8 Determining the durability indication on labels for a particular product involves considerations of microbiological safety, of retention of acceptable organoleptic properties as assessed by sensory analysis, of chemical, physical and microbiological characteristics, and, in the case of nutrition information including vitamin declaration(s), compliance with the amounts declared. It is not enough to guess at the shelf life or to use what is thought to be (or worse, longer than is thought to be) the shelf life used by another manufacturer for an apparently similar product.

37.9   Proper scientifically based testing should be carried out to establish the shelf life to be used for a product in a specific packed
format. For details of the circumstances relating to the use of
'best before' and 'best before end' dates, and the circumstances
that trigger the requirement for a 'use by' declaration, see **Directive
2000/13/EC** as amended, and in the UK the **Food Labelling
Regulations 1996, SI 1996/1499**. These regulations were revoked
by the **Food Information Regulation 2014, SI 2014 No. 1855** that
implements **EU Regulation No. 1169/2011** on the provision of
food information to consumers and **Commission Delegated
Regulation (EU) No. 78/2014** amending Annex II of EU food
information to consumers listing the 14 allergens. The **Food
Information Regulation 2014** brings together general labelling,
nutrition and allergen labelling in a single framework legislation
(see the *Food Law Guide*, which can be accessed on the Food
Standards Agency website). Some of the major requirements in
the legislation that affect food manufacturers are as follows:

- For prepacked foods, allergens must be emphasised in the ingredients list.
- A minimum font size must be applied to mandatory
information.
- Country of origin (COO) information is required for fresh
and chilled meat from swine, sheep, goats and poultry, implementing **EU Regulation No. 1337/2013**, which lays down rules
for this requirement.
- Added water over 5% must be declared in the name of the
food for meat products and meat preparations that have the
appearance of a cut, joint, slice, portion or carcase of meat.
This also applies to fish products that have the appearance of
a cut, joint, slice, portion, fillet or whole fishery product.
- If a manufacturer produces food for sale at a stage prior to
the final consumer, i.e. through a wholesaler or to a caterer or
other third party, it is the responsibility of the manufacturer
to ensure that information on all the mandatory particulars is
provided. Not all the information will be required on a label
and some information may be provided on commercial documents either accompanying the food or sent prior to delivery.[1]
The mandatory particulars outlined in Article 9 of the **EU
Regulation No. 1169/2011** are as follows:
    (a) the name of the food;
    (b) the list of ingredients;
    (c) any ingredient or processing aid listed in Annex II or
    derived from a substance or product listed in Annex II
    causing allergies or intolerances used in the manufacture
    or preparation of a food and still present in the finished
    product, even if in an altered form;
    (d) the quantity of certain ingredients or categories of
    ingredients;
    (e) the net quantity of the food in the appropriate unites (e.g.
    kg/g/mg or l/cl/ml);
    (f) the date of minimum durability or the 'use by' date;

(g) any special storage conditions and/or conditions of use;

(h) the name or business name and address of the food business operator;

(i) the country of origin or place of provenance where required;

(j) instructions for use where it would be difficult to make appropriate use of the food in the absence of such instructions (if food needs to be cooked the label must be clear in this regard, particularly if the food requires the use of a specific technique, e.g. oven or microwave);

(k) with respect to beverages containing more than 1.2 % by volume of alcohol, the actual alcoholic strength by volume;

(l) a nutrition declaration.

37.10 The Quantitative Ingredient Declarations (QUID) labelling requirement applies to the labelling and marking of prepacked food that is sold direct to the consumer/customer in its current packaging. The QUID requirement designates the circumstances in which the percentage(s) of one or more ingredients in a product must be declared in the ingredients list. **Regulation EU No. 1169/2011** on the provision of information to consumers was published in October 2011.[1] The list of ingredients must be headed or preceded by a suitable heading that consists of or includes the word 'ingredients'. In general, all ingredients must be listed in descending order of weight as recorded at the time of their use in the manufacture of the food.[2]

Any engineered nanomaterial used as an ingredient in a food should be clearly indicated and must have 'nano' in brackets after its name in the ingredient list.

Additives must be stated in the ingredients list by their functional class, such as antioxidant or preservative, along with either their name as referred to in the additive legislation (see **EC Regulation No. 1333/2008**) or the E number.[1]

Allergenic ingredients should be emphasised in the ingredients lists by using a font, style or colour that ensures they are clearly distinguished from the other ingredients present.[1]

37.11 See Chapter 49 regarding labelling of irradiated foods, Chapter 50 regarding labelling of novel foods and genetically modified foods, and Chapter 8 regarding labelling requirements in respect of the presence of legally specified food allergens.

[1] https://www.food.gov.uk/sites/default/files/fir-guidance2014.pdf.
[2] http://eur-lex.europa.eu/LexUriServ/LexUriServ.do?uri = OJ:L:2011:304:0018:0063:EN:PDF.

38   GOOD CONTROL LABORATORY PRACTICE AND USE OF OUTSIDE LABORATORY SERVICES

Principle

A control laboratory should have appropriate premises, facilities and staff, and be so organised as to enable it to provide an effective service at all relevant times necessary to fulfil good manufacturing practice (GMP) requirements.

EU GLP

38.1   The Organisation for Economic Co-operation and Development (OECD) Principles of Good Laboratory Practice (GLP) ensure that test data is reliable and avoid duplicate testing in the wider context of harmonising testing procedures for the mutual acceptance of data. The European Union (EU) has adopted these principles and the revised OECD Guides for Compliance Monitoring Procedures for GLP as annexes to the two GLP Directives. **EC Directive 2004/9/EC** lays down the requirements for the inspection and verification of good laboratory practice and **2004/10/EC** lays down the requirements for the harmonisation of laws, regulations and administrative provisions relating to the application of good laboratory practice and the verification of their applications for tests on chemical substances. The OEDC defines GLP as 'a quality system concerned with the organisational process and the conditions under which non-clinical health and environmental safety studies are planned, performed, monitored recorded, archived and reported'.

Resources

38.2   The resources required will depend on the nature of the materials and/or products to be tested. It is essential that facilities are appropriate to the needs of the tests, whether chemical, physical, biological or microbiological. Staff should be properly trained, well motivated and well managed. Standards should be set at the highest level and maintained by careful attention to approved and agreed methods and method checks using, where appropriate, reliable outside expertise.

Premises

38.3   Control laboratories should be constructed, located, designed, equipped, maintained and of sufficient size with enough space to suit the operations to be performed in them. This will include provision for writing and recording of results and archive facilities for the storage of documents, facilities for ambient samples and refrigerated storage for samples, as required. Storage rooms should be separated from rooms where testing is undertaken and there should be adequate protection against contamination, material deterioration and pest infestation. Storage rooms should be adequate to preserve sample and reference material identity, purity and stability and the appropriate storage of hazardous materials (see 38.7). Archive facilities must ensure that the

Food & Drink – Good Manufacturing Practice: A Guide to its Responsible Management, Seventh Edition.
The Institute of Food Science & Technology Trust Fund.
© 2018 John Wiley & Sons Ltd. Published 2018 by John Wiley & Sons Ltd.

retrieval of materials and documents is timely and the contents are protected from deterioration. Storage time for documents and records needs to be defined (see Chapter 13).

To avoid contamination or loss of control of samples and test materials there should be separate areas for receipt and storage of test and reference items and it is important to check on receipt that there has been no mixing or contamination of test and reference materials.

38.4 Chemical, biological and microbiological laboratories should be separated from each other and from manufacturing areas. Separate rooms may be necessary to protect sensitive instruments from vibration, electrical interference, humidity and so on. Care should be taken to avoid contamination in either direction between laboratories (particularly microbiological laboratories, where access and exit controls should be strictly followed) and manufacturing areas, and reagents or materials that could cause taint should ideally be kept in a separate building. Provision should be made for the safe storage of waste materials awaiting disposal.

38.5 All services should be identified using colour coding according to standard documented procedures.

38.6 The UK legislation concerned with fire and explosion in laboratories falls under the process fire precautions, i.e. additional precautions that relate to workplaces such as laboratories. Laboratory practices need to comply with the **Dangerous Substances and Explosive Atmospheres Regulations 2002 (DSEAR)** and other relevant legislation as amended.

COSHH 38.7 In the UK, the **Control of Substances Hazardous to Health (COSHH) Regulations 1994, SI 1994 No 3246** as amended by **SI 1996 No 3138 Amended 2002, 2003 and 2004SI 2004/3386** affect the choice of safe laboratory working methods. All methods written up should include an assessment of the hazard of each of the chemicals used in the analysis and appropriate instructions to contain any hazard. If necessary, monitoring of the exposure to hazardous chemicals should be carried out. Advice on COSHH may be obtained (in the UK) from the Health and Safety Executive (HSE), among others.

Equipment 38.8 Control laboratory equipment and instrumentation should be appropriate to the testing procedures undertaken. Equipment must be suitably located, and have adequate capacity and appropriate design for the tests to be performed.

38.9 Equipment and instruments should be inspected, cleaned, maintained, serviced and calibrated at suitable specified intervals by an assigned competent person, persons or organisation in line with standard operating procedures. Calibration should, where appropriate, be traceable to national or international standards of measurement. Laboratories working to traceable national or international standards should in turn calibrate measuring inspection and test pieces used in the calibration process at intervals determined by risk assessment. Records of the calibration procedure and results should be maintained for each instrument or item of equipment or test piece. These records should specify the date when the next calibration or service is due (see Chapter 34).

38.10 Written operating instructions should be readily available for each instrument.

38.11 Where practicable, suitable arrangements should be made to indicate failure of equipment or services to equipment. Defective equipment should be withdrawn from use until the fault has been rectified. Following identification of defective or broken equipment, a review should be undertaken by the quality control manager or designate to determine the potential impact of equipment failure on product safety, legality or quality. Actions taken as a result of the review should be documented and verified to determine their effectiveness.

38.12 As necessary, analytical methods should include a step to verify that the equipment is functioning satisfactorily.

Cleanliness   38.13 Laboratories and equipment should be kept clean in accordance with written cleaning schedules. Adherence to cleaning schedules and compliance with cleaning standards should be verified at routine intervals. Verification activities should be recorded.

38.14 At all times personnel should wear clean protective clothing appropriate to the duties being performed, especially eye protection.

38.15 The disposal of waste material should be carefully and responsibly undertaken by appropriate bodies and in compliance with legislation. Waste must be stored in appropriate, designated areas so that the integrity of testing is not compromised.

Reagents, Controls   38.16 Where necessary, reagents should be dated upon receipt or
and Standards   preparation.

38.17 Reagents made up in the laboratory should be prepared by persons competent to do so, following laid down procedures. As applicable, labelling should indicate the identity, concentration, standardisation factor, shelf life (i.e. an expiry date) and storage instructions. The label should be initialled or signed, and dated, by the person preparing the reagent. A date for restandardisation should be recorded if required.

38.18   In certain cases, it may be necessary to carry out tests to confirm that the reagent is suitable for the purpose for which it is to be used. A record of these tests should be maintained.

38.19   Both positive and negative controls should be applied to verify the suitability of microbiological culture media. The size of the inoculums used in positive controls should be appropriate to the sensitivity required.

38.20   Reference standards, and any secondary standards prepared from them, should be dated, and be stored, handled and used so as not to prejudice their quality.

Sampling   38.21   Samples should be taken in such a manner that they are representative of the batches of material from which they are taken, in accordance with written sampling procedures approved by the quality control manager. The sampling activities used are constrained by:

• the resources and time available;
• the planned frequency of verification activities;
• the volume of data to be assessed;
• any planned or unplanned sampling bias; and
• the potential for deviation from the scope of the sampling protocol.

Sampling procedures should identify whether they are risk-based or non-risk-based methods of sampling, that is, whether all batches have an equal probability of being sampled or whether, through a risk assessment process using screening criteria, the sampling is biased towards certain ingredients, products and potential food safety hazards. Consideration needs to be given to the homogeneity of the sample or of the criteria being tested within the food product as, for example, there may be more contamination on the surface of the food product than in the sample as a whole.

For example, materials associated with possible contamination such as mycotoxins (e.g. nuts) might be sampled at a higher frequency than those products that are not identified as having the potential to be so contaminated. Risk-based sampling accepts the premise that resources are limited and that sampling must be economically viable and deliver benefits to the manufacturing business. Other screening criteria might be third-party certification, provision of certificates of analysis, history of conformance with specifications and the volume of product utilised by the manufacturing business and thus the impact of failure.

The European Commission (1976[1] and 2006[2]) distinguished between different types of food samples:

- *sampled portion:* a quantity of product constituting 'a unit' having characteristics presumed to be uniform;
- *incremental sample:* a quantity taken from one point in the sampled portion, lot or sublot; and
- *aggregate sample:* an aggregate of incremental samples taken from the same sampled portion or the combined total of all the incremental samples taken from the lot or sublot.

It is therefore important to classify the type of sample being derived as this will impact on the results in terms of the amount of variability in the batches to be sampled and whether this is masked by an aggregate sample and whether contamination in one batch is 'diluted' by the use of aggregate sampling to a point where it is not detected or finally where traceability of a food safety hazard identified during sampling to a particular batch is no longer possible. The physical state of the material, that is, liquid, solid or particulates, will also influence the sampling protocols that can be effectively adopted. Sampling procedures should include:

(a) the method and rate of sampling, which should reflect the degree of homogeneity of the material being sampled;
(b) the equipment to be used, which must also be recorded on any testing records;
(c) the amount of sample to be taken;
(d) the instructions for any required subdivision of the sample;
(e) the type and condition of the sample container to be used;
(f) the storage requirements for the sample prior to testing;
(g) any special precautions to be observed, especially in regard to sterile sampling or sampling of noxious materials and the prevention of false positives; and
(h) the cleaning and storage of sampling equipment.

Any sampling by production personnel should only be done by competent personnel in accordance with these approved procedures. Training records should be maintained to demonstrate that staff are competent.

38.22 Each sample container should bear a secure and indelible label indicating its contents, with the batch or lot number reference and the date of sampling. It should also be possible to identify the bulk containers from which samples have been drawn.

[1]European Commission (1976). First Commission Directive of 1 March 1976 establishing Community methods of sampling for the official control of feeding stuffs, 76/371/EEC. *Official Journal*, L 102, 15/04/1976, pp 1–7.
[2]European Commission (2006). Commission Regulation of 23 February 2006 laying down the methods of sampling and analysis for the official control of the levels of mycotoxins in foodstuffs, 2006/401/EC. *Official Journal*, L 70, 09/03/2006, pp 12–34.

38.23 Sampling equipment should be cleaned after each use and stored separately from other laboratory equipment.

38.24 Care should be taken to avoid contamination or causing deterioration whenever a material or product is sampled. Special care is necessary when resealing sampled containers to prevent damage to, or contamination of or by, the contents.

Methods 38.25 Methods should be chosen with care to fulfil the needs of the analyses. For quality control purposes, the chosen method should be that most efficacious for the accuracy and speed of results needed, and the skill of the staff concerned. When possible, methods acceptable to any enforcing authority, or which are internationally acceptable, should be used. In all cases, method checks need to be incorporated into any analytical scheme to ensure reproducibility, repeatability and operator independence. Reviews of the methods used should be undertaken at predetermined intervals or at times appropriate to a developed need.

Documentation 38.26 All test facilities should have a documented quality assurance programme to assure that testing undertaken is performed in compliance with the principles of GLP. An individual should be designated by the manufacturer to ensure that the quality assurance programme for testing is consistently complied with. Laboratory documentation should be in line with the general guidance given in Chapter 13. When electronic or magnetic recording methods are used, see also Chapter 39.

38.27 Retention samples should be regarded as part of the laboratory records.

38.28 It is useful to record test results in a manner that will facilitate comparative reviews of those results and the detection of trends. To assist this process, 'commodity files' may be established.

Records of Analysis 38.29 Details to be recorded on the receipt and testing of starting materials, packaging materials and intermediate, bulk and finished products are indicated in Chapter 13. Analytical records should contain:

(a) the name of the product or material and its code reference;
(b) the date of receipt and sampling;
(c) the source of the product or material (including supplier and country of origin);
(d) the date of testing;
(e) the batch or lot number;
(f) an indication of tests performed;
(g) reference to the methods used;
(h) results;
(i) any decision regarding release, rejection or other status; and
(j) the signature or initials of the analyst, and the signature of the person taking the above decisions.

38.30 In addition to the above records, analysts' laboratory records of the basic data and calculations from which the test results were derived should also be retained (e.g. weighing, readings, recorder charts).

Specifications   38.31 Specifications approved by quality control and including analytical parameters should be established for all raw materials and bulk, intermediate and finished products (see Chapter 13).

Testing   38.32 The persons responsible for laboratory management should ensure that suitable test methods, validated in the context of available facilities and equipment, are adopted or developed.

38.33 Samples should be tested in accordance with the test methods referred to, or detailed, in the relevant specifications. The validity of the results obtained should be checked (and, as necessary, any calculations checked) before the material is released or rejected.

38.34 In-process control work carried out by production staff should proceed in accordance with methods approved by the person responsible for quality control.

38.35 Microbiological testing should be carried out in appropriate facilities and with due consideration to the handling, storage and disposal of samples and tested materials.

Contract Analysis   38.36 Although analysis and testing may be undertaken by a contract analyst, the responsibility for quality control cannot be delegated to him/her.

38.37 The nature and extent of any contract analysis to be undertaken should be agreed and clearly defined in writing, and procedures for taking samples should be set out as discussed in 38.21–38.24.

38.38 The contract analyst should be supplied with full details of the test method(s) relevant to the material under examination. These will need to be confirmed as suitable for use in the context of the contract laboratory.

38.39 Formal arrangements should be made for the retention of samples and of records of test results. Protocols must be in place for timely reporting of the results that identify product non-conformance for product safety, legality and quality to the quality control manager or designate.

Accreditation   38.40 Where control laboratories are located within factories that are applying for third-party certification, the laboratory is likely to be considered as part of the quality control system assessment. Adherence to GLP should meet the technical standards demanded, although the documentation of the system will need to conform to the needs of the assessors. In the UK, many food laboratories will wish to seek accreditation under the UK Accreditation Service (UKAS) accreditation scheme. Third-party laboratories used for testing should be accredited by a competent national authority, such as UKAS, to ISO/IEC 17025 or an equivalent standard. This is often a requirement for third-party certification of a manufacturer. Approval may be withdrawn from any laboratory if standards in any of the accreditation schemes are not maintained. Proficiency testing schemes are available for laboratories to monitor and give confidence in laboratory methods. Participating laboratories will take part in the Food Analysis Performance Assessment Scheme (FAPAS) testing (see https://fapas.com for further details).

39 ELECTRONIC DATA PROCESSING AND CONTROL SYSTEMS

Food & Drink – Good Manufacturing Practice: A Guide to its Responsible Management, Seventh Edition.
The Institute of Food Science & Technology Trust Fund.
© 2018 John Wiley & Sons Ltd. Published 2018 by John Wiley & Sons Ltd.

Principle

The use of electronic data processing, including the use of computers for storing and handling any kind of documentation referred to in Chapter 13, or the use of computers or electronic devices in control of processes or operations, does not in any way alter the need to observe the relevant principles set out in this Guide. However, it imposes the additional need for safeguards to ensure that there is no consequent adverse effect on the achievement of requisite product safety, legality, integrity or quality, and no risk of required records or data being irretrievably lost, damaged, corrupted or altered in other than a properly authorised way.

Implementation and Operation

39.1 Before a computer system is put into operation, the required purpose should be clearly defined, and it should be tested for capability to achieve that purpose. If a manual system is being replaced, or an existing system is being replaced by a new one, the old system should be continued alongside for a period, both as part of the validation of the new system and as a safeguard in the event of problems or teething troubles in the new system.

39.2 Access to data and documents should be readily obtainable by authorised persons where necessary. The risk of unauthorised access to computer systems should be considered as part of the threat analysis critical control point (TACCP) risk assessment (see Chapters 5, 6 and 7).

39.3 An effective back-up system is essential (see 39.10).

Responsibility for Systems

39.4 The use of computers and electronic data control systems does not change the responsibilities, as set out elsewhere in this Guide, of key personnel but additionally requires close collaboration between them and those responsible for the computer systems. Where this responsibility lies should be clearly stated. Persons using an electronic system should be appropriately trained in its use, and should be able to obtain rapid expert advice to deal with any problems that might arise in its use. The requirements outlined in Chapter 13 with regard to document and data control apply equally to electronic data, as they do to data in paper format. This is especially so for the management of obsolete files, documents and data held electronically either in a memory that has personal access only or on a server. Consideration should be given to the use of read-only files by those who can access but do not have authority to alter or amend electronic documents such as policies, procedures and work instructions.

39.5 **Where the system contains personal information (e.g. related to consumer complaints), governed by the UK Data Protection Act 1998, the requirements of the Act and as amended must be observed**.

39.6 Alterations to the system or a computer programme should be made only with proper authorisation and in accordance with a defined procedure, which should include the checking, approving and implementing of the change. When any such change is made, previously stored data should be checked for accessibility and accuracy. A log of all such alterations should be maintained to record the date, the changed details, the person making the alteration, authorisation details and any other pertinent information.

39.7 By means of appropriate keys, pass cards, personal codes or passwords and restriction of access to computer terminals and recording media, the system should contain safeguards against unauthorised access to, or alteration of, any data (see Chapter 7). Personnel, visitor and contractor procedures should clearly outline the protocols that are in place for the ownership and use within the organisation of portable back-up hard drives, memory sticks and pens, laptops, tablets or other mobile electronic devices that are brought onto the premises. Where the computer is part of a network (whether internal or external), security measures against unauthorised access should be taken. Where the same equipment is used for other functions (e.g. accounting, sales records, personnel records), access to the types of data referred to in this chapter should not enable access to data pertaining to such other company functions. Data should only be obtained, entered or amended by persons authorised to do so, and there should be a defined procedure for the issue, cancellation or alteration of authorisation. Where critical data are altered by an authorised person, the record should show that the alteration has been made, by whom and the reason.

39.8 When critical performance data are being entered on a computer file record, there should be an independent check on the accuracy of the entry.

39.9 The system should be designed, precautions provided and controls (of the type mentioned in Section 39.10) exercised so as to safeguard against accidental or wilful damage of stored data by persons or physical or electronic means. A system of cross-checking for any loss of data is advisable.

Back-up 39.10 There should be available adequate back-up arrangements for any system that needs to be operated in the event of a breakdown or emergency, and such back-up arrangements should be capable of being called into use at short notice. As a safeguard against loss or corruption of stored data and documents, provision should be made for back-up copies of data and systems software to be stored remotely from the computer's location; such copies should be kept up to date and tested periodically. Access to the types of documents listed in Chapter 13 should be available to the quality control manager and the production manager.

39.11 The procedures to be followed in the event of a system failure or breakdown should be defined and routinely tested. Any failures and remedial action taken should be recorded.

39.12 If hardware service or software maintenance is provided by an outside agency, there should be a formal agreement that includes a clear statement of the responsibilities of that agency, including a clause on the maintenance of confidentiality.

39.13 Consideration should be given to the ease of access to files with regard to system hacking, viral attack or other activity that could compromise the security of controlled data.

40 SUSTAINABILITY ISSUES

Principle

In the Introduction, the intention was made clear to limit this Guide to matters having a direct bearing on the scientific, technological and organisational aspects affecting the quality, legality and safety of products. For this reason, detailed consideration has not been given to sustainability as an issue, but supply chain requirements for the manufacturer to demonstrate that they are sustainable have an impact on the factory and its operations.

General

40.1 For many manufacturing businesses it is a prerequisite to supply that they can demonstrate that they have a business sustainability plan in place. This often forms part of an overall corporate social responsibility (CSR) strategy. CSR considers the role that organisations play in how they act and react to stakeholder concerns over environmental and social issues. Whilst the bottom line is often used as a term to describe profitability, sustainability considers three elements: economic sustainability, environmental sustainability and social sustainability, the 'triple bottom line'. Personnel, responsibilities and training (Chapter 17), worker welfare standards (Chapter 18), waste management (Chapter 30), food donation controls and animal food supply (Chapter 31) and environmental issues (Chapter 41) are considered in this Guide and demonstrate some of the elements of good manufacturing practice that influence organisational sustainability.

40.2 Extrinsic product characteristics such as animal welfare standards, social standards or environmental standards are often used to communicate to consumers the way that ingredients or manufactured products have been grown, so-called business-to-consumer (B2C) standards, for example Fairtrade and organic standards. Certification to these standards can result in logos being placed on product packaging to communicate directly to consumers. It is important that provenance is maintained during all manufacturing processes so integrity can be demonstrated (Chapter 15), especially in the event of reworked product (see Chapter 29). Certification to third-party standards is also a way for manufacturing organisations to demonstrate their sustainability credentials to their customers through business-to-business (B2B) standards. These standards include worker welfare standards (see Chapter 18), SA8000, the social accountability standard, and ISO 26000 guidance on social responsibility. ISO 26000 provides guidance on how organisations can operate ethically, contributing to sustainable development whilst also taking into account legislative requirements, stakeholder expectations and accepted business behaviour. The ISEAL Alliance[1] has a range of information on sustainable sourcing standards.

[1] https://www.isealalliance.org.

Food & Drink – Good Manufacturing Practice: A Guide to its Responsible Management, Seventh Edition.
The Institute of Food Science & Technology Trust Fund.
© 2018 John Wiley & Sons Ltd. Published 2018 by John Wiley & Sons Ltd.

40.3   CSR performance is often measured through the development of key performance indicators or metrics. Corporate social performance, sustainability indexes and social responsibility disclosure frameworks such as the Global Reporting Initiative encourage more companies to undertake sustainability disclosure. As outlined in 40.2 disclosure can be either B2B or B2C and the extent to which indexes such as the Dow Jones Sustainability Index and FTSE4Good are used is growing.

Principle

In the Introduction, the intention was made clear to limit this Guide to matters having a direct bearing on the scientific, technological and organisational aspects affecting the quality, legality and safety of products. For this reason, detailed consideration has not been given to the impact of the factory and its operations on the external environment. It is, however, acknowledged here that the management of any food manufacturing operation has general responsibility and, in most countries, legal obligations (with which it must be familiar) for these aspects.

General

41.1   The premises, equipment, personnel, manufacturing operations, intake of materials, despatch of products and treatment/disposal of unwanted by-products (waste materials, by-products unsuitable for human consumption, effluent, emissions of smoke, gases, fumes, dust, noise, light and odours that are offensive or that may cause taint elsewhere) must be controlled and comply with:

(a) all the appropriate environmental legislation in the country concerned;

(b) any specific legislation in the country in which manufacture takes place relating to the treatment and handling of specific by-products; and

(c) the environmental legislation in the country in which manufacture takes place as well as any additional requirements of the local authority, and should have full regard to the responsibility of industry to take whatever steps are necessary to minimise, and preferably eliminate, any adverse environmental impact of its operations.

41.2   Environmental management is just one element of a business sustainability plan for a manufacturing business. The so-called 'triple bottom line' of a food business includes social (people), environmental (planet) and economic (profit) dimensions. This is discussed more fully in Chapter 40.

41.3   When building food manufacturing premises, consideration should be given to the local environment and the actions that may need to be taken to protect the environment. This is usually addressed by undertaking an environmental risk assessment that once developed and implemented must be routinely reviewed and updated as activities and practices on the site change. The factors to be considered include, but are not limited to:

(a) local activities that could impact on food production;

(b) chemical storage on-site and chemical mixing areas;

(c) fuel and oil storage on-site;

(d) potential for existing ground contamination or pollution from spillages;

(e) design of drainage systems, including storm water and designated waste drains;

(f) sources of and conservation of water and energy;

(g) waste-water treatment and the need for effluent treatment;

(h) storage areas for packaging awaiting recycling that minimise harbourage for pests; and

(i) storage of other wastes awaiting collection, for example in a poultry meat factory the storage of blood, sludge or feathers.

41.4 Formal policies and protocols such as environmental policies, waste management and waste minimisation protocols (including recycling policies; see Chapter 30), and dust, odour and noise management plans need to be developed, implemented and periodically reviewed to ensure that they are still valid, are being complied with and are effective. Resource management procedures should also be developed to effectively manage resources utilised within the manufacturing unit, for example water, energy and materials.

41.5 Manufacturing organisations may seek third-party certification of their environmental management systems for compliance with standards such as EN ISO 14001:2015.

Environmental Permitting

41.6 Integrated pollution prevention and control (IPPC) was operated under the **Pollution Prevention and Control (England and Wales) Regulations 2000**, and similar regulations for Scotland and Northern Ireland were made under the **Pollution Prevention and Control Act 1999**, which implements the **European Community (EC) Directive 96/61/EC on IPPC**. The IPPC Directive was codified (**Directive 2008/1/EC**). The **IPPC Directive (96/61/EC)** was implemented in England and Wales through the **Pollution Prevention and Control (England and Wales) Regulations 2000**, which have been replaced by the **Environmental Permitting Regulations 2007**. **Directive 2010/75/EU** of the European Parliament and of the Council of 24 November 2010 on industrial emissions (integrated pollution prevention and control) superseded previous regulations.[1]

The scope of the legislation includes all 'installations' treating and processing materials intended for the production of food from:

• operating slaughterhouses with a carcass production capacity greater than 50 tonnes per day;

• animal raw materials (other than exclusively milk) at a plant with a finished product production capacity greater than 75 tonnes per day;

[1] http://eur-lex.europa.eu/legal-content/EN/TXT/PDF/?uri=CELEX:32010L0075&from=EN.

- vegetable raw materials at a plant with a finished product production capacity greater than 300 tonnes per day or 600 tonnes per day where the installation operates for a period of no more than 90 consecutive days in any year;
- animal and vegetable raw materials, in both combined and separate products, with a finished product production capacity in tonnes per day greater than 75 tonnes if A is equal to 10 or more or [300 − (22.5 × A)] in any other case, where A is the portion of animal material (percentage of weight) of the finished product production capacity.

Packaging should not be included in any calculations of the final weight of the product.

Activities in this sector include production and preserving of meat, fish and potatoes, manufacture of fruit and vegetable juice, fruit and vegetable processing, milk processing, cereal processing and production or processing of animal feed, pet food, bread, cakes and biscuits, sugar, chocolate and confectionery, pasta products, tea and coffee, and beverages and brewing.

This legislation requires manufacturing sites to identify their impact on air, soil and water using an integrated approach in order to develop environmental management systems to minimise the impact of the site. This could include the development of documented management systems to meet the needs of EN ISO 14001:2015. Consideration should also be given to the potential for activities and practices undertaken at the manufacturing site impacting on local environments and neighbours, especially those sites that are environmentally sensitive.

42   HEALTH AND SAFETY ISSUES

Principle

In the Introduction, the intention was made clear to limit this Guide to matters having a direct bearing on the scientific, technological and organisational aspects affecting the quality and safety of products. For this reason, detailed consideration has not been given to the safety and welfare of operatives (except insofar as their health and personal hygiene bears directly on the quality and safety of products). It is, however, acknowledged here that the management of any food manufacturing operation has general responsibility and, in most countries, legal obligations (with which it must be familiar) for the health and safety of its employees.

General

42.1   In the UK, the **Health and Safety at Work Act 1974** is the primary piece of legislation covering occupational health and safety. The Act applies not only to employees but also to members of the public, including customers, visitors and contractors. A duty is placed on employers with more than five employees to develop and implement a written health and safety policy. Statutory instruments are secondary pieces of legislation made under specific Acts of Parliament that address specific health and safety issues. The range of health and safety legislation that food manufacturers must comply with is too complex to outline in this chapter. However, food manufacturers must have a formal, appropriate and effective health and safety management system that ensures health and safety issues are minimised. For comprehensive support with regard to the requirements for health and safety management within a manufacturing unit see the Health and Safety Executive (HSE) website (www.hse.gov.uk).

42.2   The main causes of injury in a food manufacturing environment include manual handling injuries, slips on wet or food-contaminated floors, falls from heights, workplace injuries including those involving fork-lift trucks, injuries associated with food processing machinery, packaging machinery and being hit by a falling object, for example from racking, contact with harmful substances, or being injured by tools used, such as knives. The risks associated with a given food manufacturing environment are specific to the foods being processed and the tasks being undertaken, for example the use of ovens and heat-related injuries or risk of dust explosions is very specific to one type of food manufacture and not another. The food manufacturer therefore needs to consider all potential health and safety hazards and undertake a formal risk assessment to ensure that these risks are managed effectively and the risk of minor and major injury is reduced.

Food & Drink – Good Manufacturing Practice: A Guide to its Responsible Management, Seventh Edition.
The Institute of Food Science & Technology Trust Fund.
© 2018 John Wiley & Sons Ltd. Published 2018 by John Wiley & Sons Ltd.

42.3 The main occupational health issues in a food manufacturing environment are musculoskeletal injuries, including back injuries, dermatitis from repeated handwashing or contact with certain foodstuffs, excessive noise that affects hearing, i.e. where noise levels exceed 85 dB(A), occupational asthma, for example from dust, occupational dermatitis and rhinitis, for example from dust, spices or seasonings, and work-related stress. **The Reporting of Injuries, Diseases and Dangerous Occurrences Regulations (RIDDOR) 2013** require employers, the self-employed and people in control of work premises (identified as the 'responsible person') to report certain serious occupational diseases, specified dangerous occurrences (near misses) and serious workplace accidents, for example a fatality to the Health and Safety Executive within a prescribed timescale from the date of the incident. Specified incidents that must be reported, known as RIDDOR, include burns, fractures, amputations, and any injury leading to blindness, loss of consciousness or admittance to hospital. For a full list of injuries and occupational diseases that are subject to RIDDOR see http://www.hse.gov.uk/riddor/reportable-incidents.htm.

42.4 In the UK, the **Control of Substances Hazardous to Health (COSHH) Regulations 1999, amended in 2002**, introduced a requirement for employers to carry out a formal assessment of all work that is liable to expose employees to hazardous substances, including chemicals, liquids, solids, vapours, fumes, mists, dusts, nanotechnology and biological agents, including microorganisms. The formal COSHH assessment must evaluate the risks to health and the actions that are required to minimise that risk. More information is available from the HSE website.

42.5 A responsible person must be appointed by the manufacturer to have overall responsibility for the health and safety management system. They are responsible for the operation of an effective health and safety management system and for implementing the health and safety strategy drawn together by senior management. Senior management are responsible for providing adequate resources to ensure that the health and safety management system can be effectively implemented and that the safety of workers is assured. The responsible person and other key individuals identified in the health and safety management system must be competent and have had appropriate training and support to ensure that they have the knowledge and skills required so that they can fulfil their responsibilities. The training and qualifications must be of sufficient quality and at a sufficient level that is commensurate with the individual's responsibilities, for example the Institute of Occupational Safety and Health suite of qualifications. Documentation should be developed in line with the general guidance given in Chapter 13. When electronic or magnetic recording methods are used, see also Chapter 39.

PART II – SUPPLEMENTARY GUIDANCE ON SOME SPECIFIC PRODUCTION CATEGORIES

This part of the Guide is not intended to be a comprehensive commentary covering every product group in detail; it expands on general principles where it has been felt that there is a specific need to do so.

Food & Drink – Good Manufacturing Practice: A Guide to its Responsible Management, Seventh Edition.
The Institute of Food Science & Technology Trust Fund.
© 2018 John Wiley & Sons Ltd. Published 2018 by John Wiley & Sons Ltd.

43   HEAT-PRESERVED FOODS

Principle

The use of heat treatment as a method of food preservation has ancient roots. The aim of using heat is to eliminate, minimise or inhibit micro-organism and enzymic activity that, if it was allowed to proliferate, would make the food unsafe or otherwise unfit for human consumption. The term 'thermal preservation' denotes the use of heat over a prescribed period of time to render a food suitable for consumption and is a means to extend product durability and shelf life. Examples of heat preservation include canning, cooking, boiling, pasteurising, sterilising and blanching amongst others. However, it should be noted that whilst some heat preservation can address vegetative bacteria a given process may not affect thermophilic micro-organisms, heat-tolerant bacterial toxins or microbiological spores. This chapter is not written as a comprehensive guide to heat preservation, but instead to give a general outline of the subject of heat-preserved foods in the context of food manufacture.

General

43.1   Heat treatment is a method of preservation that will prevent or delay product deterioration or spoilage and inhibit the growth of pathogenic organisms. There are several methods of heat treatment, including canning, pasteurisation, ultra-heat treatment and sterilisation.

Heat Treatment

43.2   Heat treatment processes within the manufacturing environment must be designed to ensure that the foods subject to such treatment are microbiologically stable and growth of micro-organisms is prevented under subsequent storage conditions, within the manufacturing process, and during the period in which the foods are intended to be suitable for storage and consumption.

Commercial Sterility

43.3   Commercially sterile food is food that has undergone a heat treatment that will destroy all vegetative pathogens and organisms capable of growth or causing spoilage in the food under standard storage conditions, for example low-acid canned food or aseptically packed product.

43.4   All low-acid foods having in any part of them a pH value of 4.5 or above and intended for storage under non-refrigerated conditions must be subjected to the minimum botulinum process, that is, one that will reduce the probability of survival of *Clostridium botulinum* spores by at least 12 decimal reductions, unless the formulation or water activity, or both, of the food is such that it can be demonstrated that growth of strains or forms of the organism cannot occur. The scheduled heat process required to achieve commercial sterility will be in excess of the minimum botulinum process ($F_o = 3$), as many organisms associated with the spoilage and economic loss of heat-preserved foods

Food & Drink – Good Manufacturing Practice: A Guide to its Responsible Management, Seventh Edition.
The Institute of Food Science & Technology Trust Fund.
© 2018 John Wiley & Sons Ltd. Published 2018 by John Wiley & Sons Ltd.

have greater heat resistance than *C. botulinum*. In addition to achieving commercial sterility, the scheduled heat process may be further extended for specific reasons, for example to soften the bones in canned fish or to tenderise or texturise meats.

43.5   Low-acid foods packed in acidic carrying fluids or whose natural pH value has been otherwise lowered by the controlled addition of acids, should be subjected to at least a minimum botulinum process unless it can be established that the equilibrium pH, including the pH value at the cores of particles, is less than 4.2 within 4 hours of the end of the thermal process, thus providing a safety margin for non-uniformity of acidification.

43.6   The scheduled heat process is the key document for the manufacture of all heat-processed foods. In the UK, it is addressed by a number of guidelines available from Campden BRI.[1]

43.7   The Statutory Instruments of minimum scheduled heat treatments for milk, semi-skimmed and skimmed milk, milk-based drinks and cream include the **Dairy Products (Hygiene) Regulations 1995, SI 1995 No. 1086, as amended**, for ice cream in **SI 1959 No. 734, as amended in 1962, 1963, 1982, 1985, 1990 and 1995**, and for liquid egg in the **Egg Products Regulations 1993, SI 1993 No. 1520.**

43.8   For efficient heat treatments, the physical, chemical and microbiological characteristics of each specific product need to be identified since these are critical factors that can affect the rate of heat penetration or heat distribution. These, therefore, need to be taken into account for plant design and processing regimes. Products should be defined in terms of their densities, rheological properties, water activity (a_w), pH, temperature, thermal properties, pressures, specific heats, microbial loading, gas content and corrosive properties. Electrical properties are critical when using electrical energy for sterilisation. The processing parameters set must be validated to demonstrate that they are capable of delivering safe food (see 43.26 and Chapter 3).

43.9   Discrete particles should be defined additionally in terms of size, shape, concentration and swelling or shrinkage. Dry ingredients should be thoroughly dispersed and wetted. Non-condensable gases should be disengaged.

43.10   Scheduled heat processes are derived by measuring the rate of heat penetration into the product and integrating the time/temperature exposure, hence determining the total lethal heat that the product receives within a particular process regime.

[1] http://www.campden.co.uk/publications/pubs.php.

For low-acid foods, or foods otherwise designated as requiring at least a minimal botulinum process, 'worst-case' assumptions involving the lowest temperature reached and shortest retention time at it should be made for *C. botulinum* spore destruction kinetics (see 43.4).

43.11 All new products or changes in the manufacturing operations, formulation and, if appropriate, containers for existing packs should be fully evaluated as to their effect on the rate of heat transfer through the product before commercial production is undertaken. Effective validation must be undertaken to ensure that such modifications, and the associated changes to processing conditions that are required, are effective. This activity must be recorded and records maintained.

43.12 Products not given at least a minimum botulinum process, or that are not otherwise shelf stable at ambient temperature, should carry clear instructions on storage conditions, including maximum storage temperatures.

43.13 All containers should be indelibly marked with a code indicating at least the place and date of production. Further information on the time of production and the production line can be very valuable. Empty containers or reel stock should be checked on receipt to ensure that they comply with the agreed recommendations for the product. They should be stored and handled to prevent their becoming contaminated or damaged and the integrity of the container thereby affected. Containers conveyed into a filling machine should be clean and flawless.

43.14 All batch and continuous processing equipment should be fitted with direct reading (indicating) temperature probes and automatic time and temperature recording instruments. These should be calibrated at designated intervals (see Chapter 34). Records of calibration and processing records, including data from temperature monitoring and automatic recording devices, should be kept for at least 3 years from the date of production depending on the duration of the product or customer requirement. Monitoring may be undertaken by on-line or quality control personnel. Personnel undertaking monitoring need to be aware of the critical nature of such monitoring and the actions to be taken in the instance that there is a product or process failure identified at a product safety critical control point (see Chapter 3).

43.15 Food processing equipment should be internally clean and disinfected, correctly assembled, free from 'dead legs' and with control systems that have instruments of reproducible accuracy that are routinely calibrated. Procedures must be in place to address the actions to be taken in the event of equipment calibration failure (see Chapter 34).

43.16 Water used for manufacturing purposes, including that used in making up products or likely to come into direct contact with the product, must be of potable quality and free from any:

(a) substances in quantities likely to cause harm to health;
(b) substances at levels capable of causing accelerated internal corrosion of metallic containers and closures or causing taints; and
(c) harmful micro-organisms.

To maintain the required microbiological standards, the water should, if necessary, be chlorinated or otherwise adequately treated (see 43.32).

43.17 Mechanical unloading and handling systems should be designed so that:

(a) dryers can be located as early as possible in the conveying system;
(b) there is minimum contact between container closures and the conveyor surfaces;
(c) abuse of containers is kept to a minimum; and
(d) conveyor surfaces must be easily cleaned and disinfected as frequently as is necessary to maintain proper standards of hygiene (see Chapter 21).

43.18 Regular microbiological surveillance, including swab tests, should be made on conveyors and other contact surfaces to establish the effectiveness of sanitation programmes (see Chapter 21).

43.19 To reduce the potential for cross-contamination between high-risk and low-risk areas (HRAs and LRAs), the post-process area and its personnel should be segregated from other sections, particularly from the preparation area of the plant. In the instance of manufacturing units that process raw and cooked meats, strict segregation of personnel should be undertaken, including segregation of welfare facilities, designated staff and designated clothing, tools and equipment.

43.20 Processed containers should be loaded into clean dry cartons or other suitable outer packaging material and stored in areas specifically allocated for this purpose. Care should be taken to minimise any damage to processed containers that could compromise seal integrity.

43.21 Samples of finished packs may be incubated as part of the overall food safety and quality control programme. However, the acceptability of the manufacturing operations should not be judged solely on the results of such tests, but rather that in most situations they are forms of verification and provide cumulative and retrospective confirmation of the efficacy of the process control and hygiene operations.

43.22 It is essential to ensure product integrity after processing and handling. Suitable precautions should be taken to ensure that package structures, including seals and seams, are free from defects that would otherwise impair their effectiveness as micro-biological barriers, and that the food does not cause breakdown of the packaging materials with which it is in contact.

43.23 The products should be suitably protected for handling and stacking in storage. They should bear identification marks that are traceable to the production records derived from the manu-facturing process (see Chapter 14).

43.24 Transport may present risks to package integrity caused, for example, by tilt and vibration, in respect of which correct handling procedures should be adopted (see Chapter 32).

43.25 Every effort should be made to take into account factors that may affect the continuing integrity of the products after they leave the immediate control of the producer. This, for example, includes handling during transportation, in wholesale and retail outlets and by the consumer. Labels warning against the use of case hooks and careless use of case opening knives should be adopted.

In-container Heat Treatment

43.26 The scheduled heat process should take into account all critical factors that may affect the rate of heat transfer in the containers. It should be established by competent and properly trained personnel using accepted scientific methods. The scheduled heat process should be validated during the product and process design phased and then revalidated in the event of changes or amendments to the original process or product design. Records giving full details of how the scheduled heat process was estab-lished and confirmed as appropriate through validation should be retained permanently on file.

43.27 Product preparation and filling operations, for example setting of headspace, that may affect the integrity of the container closure or the rate of heat penetration should be carefully controlled.

43.28 The efficiency of the closing or sealing operation should be checked before processing begins and kept under constant control to ensure the integrity of the container closure or seal. Records of quality control assessments should be kept for a designated period from the date of production depending on the duration of the product and customer requirements (see Chapter 13).

43.29 Washing of filled and closed containers, should, if required, be undertaken before the heat process is given.

43.30 Temperature distribution tests should be undertaken on all processing equipment and a venting schedule established for each type of processing unit. All changes to the services of the manufacturing site, e.g. air or gas supply, water supply, etc., and the layout of such services in relation to the processing equipment or alterations in the method of loading batch retorts should be fully evaluated for their effect on the temperature distribution and adequacy of the venting schedule.

43.31 For batch retorts, there should be a method of checking that no batch of containers has bypassed the heat process.

43.32 Potable water, including water for cooling containers, should be free from harmful micro-organisms and meet minimum quality standards and present no risk of product contamination (see Chapter 20). Borehole water run directly to retorts or municipal mains water may be used, provided that is meets micro-biological and quality standards. All water used for cooling should be sufficiently chlorinated with an adequate contact time (20 minutes or more, depending on the pH) before use such that free residual chlorine (FRC) can be detected in the water after it has cooled the containers. Tests for FRC should be undertaken at a frequency derived by risk assessment, ideally the testing should be through continuous monitoring. It is not common practice to use other chemical disinfectants to sanitise water for cooling purposes, but where approved materials are contemplated for use they should achieve the same APC quality standard and have a residual present in active form after the cooling cycle. Any changes must be fully validated as part of the food safety management process (see Chapter 3).

43.33 Cooled containers should not be manually handled while still wet.

Heat Treatment Followed by Aseptic Packaging

43.34 Aseptic technology is more complex than in-container sterilisation and needs to be thoroughly understood by the processor. It involves the production of a commercially sterile hermetically sealed package of food by pre-sterilising the food, followed by cooling, filling and sealing it in sterile containers under scheduled processes, which must be applied in order to achieve and maintain commercial sterility. Assurance of attaining these standards comes from in-process controls. Retrospective quality control analysis can only indicate the degree of success or failure in maintaining the required standards. The microbiological objective is to achieve and maintain commercial sterility.

43.35 Preproduction sterilisation of all food contact surfaces should take into account any location that may be slow to attain the set temperature. Non-food contact surfaces may be chemically treated to achieve sterilisation. Chemicals that may be used include hydrogen peroxide. The integrity of the total system should be checked at this stage.

43.36 Aseptic fillers should be cleanable, sterilisable and capable of being isolated by a separating sterile barrier and with control systems that have instruments of reproducible accuracy, which are routinely calibrated. Aseptic filling machines may be affected in their performance by dust or soil in the external environment, and therefore careful consideration should be given to their location. A filling environment should have internal surfaces suitable for cleaning and microbiological sterilisation. Any inflow of sterile air or gas should have suitable flow, velocity and direction. With laminate packaging, the operation of the sealing equipment is also critical in ensuring that the packaging is adequately sealed to prevent air or bacterial ingress. Quality control personnel and/or machine operators should ensure that the machines have been set up correctly and the resultant sealing meets specified requirements before production commences.

43.37 Aseptic fillers should be appropriate for both product and packaging. Product temperature should also be compatible with the packaging material used. Control over dispense volume is critical since this may affect the integrity of the aseptic seal of the packages or the internal stress caused by the product in the finished packages.

43.38 During start-up, standby or shutdown of an aseptic filling machine, special conditions may apply, for example in the supply of decontaminated air or gas, application of heat or application of chemicals. Care must be taken to ensure that all personnel health and safety procedures are followed and any fire risk is minimised.

43.39 Prepacking sterilisation of a product needs to take into account the critical factors to be considered when determining a scheduled heat process (43.8–43.11).

43.40 In continuous-flow sterilisation, the control system should ensure that the correct product sterilisation temperature and holding time are achieved. It should be designed to fail-safe in the event of malfunction, and under-processed food should be routed out of the system and not transferred to the aseptic storage or filling section. The plate packs on pasteurisers should be routinely checked for integrity and to ensure that there are no pinholes in the plates. The relative pressure of the product and the cooling water should be reviewed to ensure that the risk of cooling water contamination through a pinhole/crack is minimised.

General Conditions 43.41 Direct heating systems may use electrical energy instead of steam. The energy and heat transfer mechanisms are different and should be understood by those responsible for the process design, the control system and its operation.

43.42 During continuous-flow product cooling there should be an absence of microbiologically contaminating leaks.

43.43 Equipment used for batch sterilisation of product should have sensors, indicators, controllers and recorders for temperature and pressure. Bulk mixing patterns are important where they relate to temperature distribution. There should be a capability of aseptic transfer of product and an absence of microbiologically contaminating leaks.

43.44 Any methods used for microbiological decontamination of packaging should be compatible with the packaging materials and also take into account the microbiological loads. The methods should be designed using proven procedures. When scheduled processes for microbiological sterilisation of packaging materials depend on synergistic effects, these are critical factors that should then be monitored.

43.45 Heat-based processes depend on suitable temperature, pressure, humidity, time of exposure and extent of air venting, and all processes should be formally validated and then revalidated at routine intervals. Monitoring and verification activities should be adopted to ensure that critical limits as well as target levels and tolerances have been complied with during processing.

43.46 Chemical applications depend on type, concentration, dose, temperature, coverage and contact time, and all processes should be formally validated and then revalidated at routine intervals. Monitoring and verification activities should be adopted to ensure that critical limits as well as target levels and tolerances have been complied with during processing.

43.47 Ultraviolet radiation has biocidal activity that is dependent on wave and band frequency, power, distance, reflectance, temperature, exposure time and age of lamp. Dust particles may have a shielding effect. Ionising radiations exert sterilising effects depending on the dose, therefore applications of this nature must be validated and then revalidated at routine intervals. Monitoring and verification activities should be adopted to ensure that critical limits as well as target levels and tolerances have been complied with during processing.

43.48 Packaging materials that are microbiologically decontaminated by the foregoing methods should be protected from recontamination.

43.49 Packaging materials may require special conditions of storage and must be stored and handled to minimise damage that can affect packaging integrity, especially with laminate packaging.

43.50 Temperatures and humidity should be controlled within acceptable limits, and additional safeguards should be applied where pre-sterilised packaging materials or packages are used.

Microbiological Criteria

43.51 **EU Regulation No. 2073/2005 on the Microbiological Criteria for Foods (as amended by EU Regulation No. 1441/2007)** applies to all UK food businesses involved in processing, manufacturing, handling and distributing food. The Regulation introduces two types of microbiological criteria:

- *Food safety criteria* are used to assess the safety of a given product or batch and apply throughout the shelf-life. Food safety criteria must be met and if they are not then the manufacturer cannot place the food on the market or if the food is already distributed the manufacturer must withdraw or recall the batch(es) involved from the market (see Chapter 27). Corrective and preventive action must then be implemented as deemed appropriate to prevent a reoccurrence and ensure that food products comply with prescribed food safety criteria in the future (see Chapter 28).
- *Process hygiene criter*ia help show that the production processes are operating correctly at designated stages of manufacturing and handling. If a process hygiene criterion is exceeded then this should lead to a review of manufacturing procedures to improve process hygiene to the levels required. The review should also include consideration of whether there is a need for the product to be reprocessed, withdrawn or recalled. Any actions taken must be recorded (see Chapter 28).

The microbiological criteria outlined above can be used by the food manufacturer to both validate and also verify the food safety management system (FSMS) that has been adopted and implemented, and to demonstrate that the food safety plan is effective (see Chapter 3).

43.52   Except for carcasses, minced meat, meat preparations and mechanically separated meat, where minimum sampling frequencies are specified in UK legislation, **EU Regulation No. 2073/2005** does not specify minimum requirements for sampling and testing nor does it require a manufacturer to undertake positive release, i.e. to wait for the test results from the product and process monitoring that has been carried out before releasing the associated product to the market. However, the food manufacturer is required to demonstrate that their FSMS is both appropriate and effective. An appropriate sampling and testing protocol must be included within the FSMS and include both food safety and process hygiene criteria. See Annex 1 of **EU Regulation No. 2073/2005** for more details.[2]

---

[2]http://eur-lex.europa.eu/legal-content/EN/TXT/PDF/?uri=CELEX:32005R2073&from=EN.

44   CHILLED FOODS

Principle

Chilled foods are perishable foods that, in order to extend the time during which they remain wholesome and safe to eat, are kept within controlled and specified conditions, normally below 8 °C. Chilled foods may be divided into those that potentially offer a low risk of the growth of pathogenic organisms and those where the risk is theoretically high.

General

44.1   The factors that invoke the need for additional care in the manufacture of chilled foods are:

(a) the perishability of raw materials;
(b) minimal processing to maximise sensory quality;
(c) the potential for spoilage and/or pathogen growth;
(d) the rapid despatch of finished products; and
(e) chilled food chain requirements.

44.2   Given the short shelf life of chilled prepared foods, the emphasis must be on the use of hazard analysis critical control point (HACCP) to develop a system to control processes, premises, personnel and the hygienic status of ingredients/raw materials used, rather than end-product testing (see Chapter 3). Product safety must be determined by proper consideration of the following:

1.   ingredient hygienic quality;
2.   product formulation/characteristics;
3.   processing parameters;
4.   allergen control;
5.   intended use of product;
6.   storage and distribution conditions;
7.   manufacturing hygiene; and
8.   shelf life.

Ingredients

44.3   Points to consider:

•   Which pathogen(s) and level of contamination might be present?
•   Is there a possibility of preformed toxins?
•   What are reasonable specification levels to apply to minimise risk?
•   What further processing is to be applied?
•   Does the shelf life of the ingredients exceed the shelf life of the finished product?

44.4   Specifications for ingredients/raw materials should include microbiological standards. Ideally, on-site facilities for microbiological

Food & Drink – Good Manufacturing Practice: A Guide to its Responsible Management, Seventh Edition.
The Institute of Food Science & Technology Trust Fund.
© 2018 John Wiley & Sons Ltd. Published 2018 by John Wiley & Sons Ltd.

testing should be available. Where this is not available, provision should be made for adequate microbiological testing at a laboratory that has suitable third-party accreditation (see Chapter 38). The UK Health Protection Agency (now Public Health England (PHE)) in England in 2009 produced guidelines for assessing the microbiological safety of ready-to-eat foods placed on the market.[1]

44.5   Perishable ingredients/raw materials should be purchased only from approved suppliers, who should furnish regular test results and agree to warn the purchaser of any problems in maintaining the standards agreed (see Chapter 23).

44.6   Deliveries of highly perishable raw materials should not be accepted if their temperatures fall outside agreed specified ranges. Guidance on recommended storage conditions should be given on outer packaging. Temperature checks undertaken must be recorded and the equipment used must be calibrated at designated intervals.

44.7   Inspection of perishable raw materials should be based on risk assessment and may include rapid indicative testing methods. However, since testing in itself is not a control measure, as recognised by HACCP, the emphasis is on proactive controls.

Product Formulation/ Characteristics   44.8   The growth of pathogenic micro-organisms can be controlled by product formulation/characteristics. This might include:

- adjustment of pH; and/or
- adjustment of water activity (a_w); and/or
- addition of chemical preservative.

An individual factor, such as pH, may be used to reduce microbiological growth. Introducing an additional factor, such as water activity, may produce a synergistic effect, that is, the combination of the two factors as hurdles reduces microbiological growth to a greater than expected or calculated extent. Relatively small changes to both hurdles together (e.g. $pH + a_w$) may be as effective as large changes to either hurdle in isolation. Chemical preservatives are rarely added direct to chilled prepared foods, but may be used in the preservation of chilled cooked meats, for example. However, where they are used, this must be in compliance with the regulatory requirements of the country concerned, and should be at the minimum effective level to ensure product safety, while not themselves presenting a food safety hazard.

[1] http://www.hpa.org.uk/webc/HPAwebFile/HPAweb_C/1259151921557.

44.9   This aspect should be considered under the following subdivisions and outcomes:

(a) *Heat Treatments*
- None or less than 70 °C for 2 minutes. Possibility that all pathogens present will survive.
- Heated to 70 °C for 2 minutes (or equivalent). All vegetative pathogens present will be reduced to an acceptable level (6 log reduction), for example *Listeria monocytogenes*, *Staphylococcus aureus*, Salmonellae and verocytotoxigenic *Escherichia coli*. However, spores and preformed toxins may persist. The potential for recontamination by vegetative pathogens must be limited by the use of a high-risk area (HRA) (see 44.17).
- Heated to 90 °C for 10 minutes (or equivalent). In addition to vegetative pathogens, spores of psychrotrophic (non-proteolytic) *Clostridium botulinum* will be reduced to an acceptable level (e.g. 6 log reduction in this case). However, more heat-resistant spores, for example strains of *Bacillus cereus* and some preformed toxins, may persist. Measures to prevent the outgrowth of psychrotrophic *C. botulinum* must be in place where chilled products have a shelf life of more than 10 days.

(b) *Cooling*
- Heated product should be cooled as quickly as possible through the so-called danger zone that is the temperature range of 5–63 °C. This will minimise risk of spore germination and outgrowth. The time taken for cooling will vary from product to product but, as a guideline, should be no more than 2 hours. Blast chillers are often used to achieve this temperature profile.

(c) *Packaging*

This should be considered under the following subdivisions and outcomes:

- Product cooked in-pack. Pathogens will be eliminated to the extent indicated in 44.9a and there will be no opportunity for recontamination that may present a food safety hazard (assuming complete integrity of the pack/seal).
- Product cooked, cooled and assembled. Pathogens will be eliminated to the extent indicated in 44.9a, but there is risk of recontamination during assembly that may present a food safety hazard.

Modified atmosphere packaging (MAP) or vacuum packaging may be used to reduce microbiological growth in conjunction with chilled storage, but will not necessarily inhibit the growth of pathogens. There may be particular concerns with respect to *C. botulinum,* which is anerobic. It is important that the effectiveness of MAP or vacuum packaging is assessed in each case, with reference as necessary to specialist advice given that most pathogens are facultative anaerobes.

Intended Use 44.10 The key point to consider is whether the product is to be eaten:
of Product

- without further heating (i.e. is ready to eat);
- following domestic reheating (i.e. requires reheating only prior to consumption); or
- following domestic cooking (i.e. requires a heat process equivalent to at least 70°C for 2 minutes prior to consumption).

On-pack instructions are an important control measure to ensure that correct procedures are followed and food safety risk is reduced. Any instructions provided by the manufacturer must be validated to demonstrate their efficacy with records retained, and in the event of a change to the product the instructions must be revalidated, for example increasing the size of chicken pieces in a ready meal or pie.

The Chill Chain 44.11 Temperature and time control are the principal controlling factors for the safety of chilled foods. Effective temperature control throughout the chill chain is particularly important to slow or inhibit the growth of pathogenic bacteria. Chilled foods, for reasons of safety or quality, are designed to be stored at refrigeration temperatures (at or below 8°C, targeting 5°C) throughout their entire life. The performance of the proposed distribution chain should be validated and monitored by the responsible party and taken into account when specifying shelf life. The number and location of temperature probes in storage should be such as to ensure effective monitoring. The manufacturer must consider whether automatic failure alert systems, such as visual or audible alarms, or processing fail-safe systems should be in place. If such systems are used, then they must be monitored at a designated frequency by either quality control or production personnel who have been adequately trained and understand the significance of such failures and the appropriate action to take.

Records should be maintained of the monitoring and verification activity and the corrective action taken in the event of failure. If these points in the process are deemed within the HACCP plan to be critical control points (CCPs), there must be an appropriate level of awareness among all personnel of the consequence of a target level and tolerance and/or a critical limit being exceeded. Visual or audible alarms can be designed into the process to be activated if a critical limit or target level is exceeded, for example at the metal detection stage and diversion of product at a pasteuriser. The automatic failure alert systems must be tested on a routine basis and records maintained of the tests, the results and any remedial action taken. Calibration procedures must be in place for all automatic failure alert systems (see Chapter 34).

Manufacturing 44.12 The purpose of establishing specified standards of personnel and
Hygiene premises hygiene is to control the hazard of microbiological contamination and cross-contamination. The level of hygiene

required will depend on the risk of and/or the consequences of cross-contamination (see Chapters 19 to 22).

44.13 Chilled foods are manufactured using a wide variety of raw materials, processes and packaging systems, and therefore must be expected to have both differing microbiological profiles after manufacture or storage and differing shelf lives, but must be microbiologically safe at the point of consumption. Thus, products made entirely from ingredients that are heated in a container or are assembled from heated ingredient(s) under the special hygiene conditions defined as 'high risk' are designed to be free from vegetative pathogens, but not all sporeformers. Those containing raw ingredients will from time to time contain vegetative pathogens such as *Listeria* as well as toxin producers and/or spore formers. This difference must be taken into account when specifying shelf life in terms of time and temperature and consumer instructions. Since *Listeria* is able to grow under chilled temperature conditions, and is responsible for the greatest number of deaths from foodborne illness in the UK, EU and USA, this is an important organism to control. In doing so, the risk of other vegetative pathogens being present is also minimised.

Shelf Life   44.14 Shelf life depends on the control of all the preceding factors, but must be validated by challenge testing and/or modelling for each product and process for defined chill storage conditions. It must be recognised that the integrity of the whole of the chill chain is vital to ensure the safety and quality of chilled foods.

44.15 By developing a product utilising a combination of factors such as raw material quality, hygienic processing, temperature, water activity, acidity and modified atmosphere, microbiological growth can be controlled and thus spoilage and/or food-borne illness prevented. The choice and combination of hurdles will determine the shelf life and the conditions of use of the products.

44.16 Consult CCFRA Guideline G46 *Evaluation of Product Shelf Life for Chilled Foods* (2004) and *Shelf life of ready to eat food in relation to L. monocytogenes—Guidance for food business operators* (CFA/BRC/FSA, 2010) for further information on the determination of shelf life. Further guidance can be found in the guidance section of this publication (Appendix IV).

Application of HACCP   44.17 There are distinct terms that need to be considered when applying HACCP to the safety of chilled foods:

Ready to eat (RTE): Food intended by the producer or the manufacturer for direct human consumption **without the need for cooking or other processing** that is effective to reduce to an acceptable level or eliminate microorganisms of concern (i.e. cold eating).

Ready to cook (RTC): Food designed to be given a heat process by the consumer that will deliver a 6-log kill with respect to vegetative pathogens (a minimum process equivalent to 70 °C for 2 minutes) throughout all components.

Ready to reheat (RTRH): Food manufactured in a high-care area (HCA) or HRA that has been designed to be reheated by the final consumer.

Low-risk area (LRA): An area where good manufacturing practice standards are in place as described within this publication, but the area and the practices have not been specifically designed to minimise microbial contamination, for example raw material intake, storage areas of RTC foods and packaged product where the product is fully enclosed (see 43.19).

High-care area (HCA): An area designed to a high standard of facility specification and hygienic design where practices relating to personnel, ingredients, equipment and environment are managed to **minimise** microbial contamination of a RTE or RTRH product containing uncooked ingredients. HCAs include:

- areas where RTE and RTRH food is being produced/assembled; and
- areas where RTE/RTRH ingredients **not** thermally processed (minimum 70 °C for 2 minutes) **but** that have been decontaminated **and** grown/produced to RTE standards are stored and handled.

High-risk area (HRA): An area designed to a high standard of facility specification and hygiene design where practices relating to personnel, ingredients, equipment and environment are managed to **minimise** microbial contamination of a RTE or RTRH product comprising only cooked ingredients:

- areas where RTE and RTRH food is being produced/assembled; and
- areas where **only** thermally processed foods (minimum 70 °C for 2 minutes for <10-day shelf life) are stored and handled.

Product safety requires the application of HACCP principles to the design of plant, product and process, the management of hygiene procedures and the control of the process. HACCP is an evolving risk management system – it must be reviewed and revalidatd when any change is made to plant, raw material specifications, process or packaging (see Chapter 3).

44.18   Figure 44.1 represents decision trees formulated by the Chilled Food Association (CFA) (http://www.chilledfood.org), whose permission to include them and assistance with the preparation of this chapter are gratefully acknowledged. The first decision tree shows the basic decision pathway. Both the heat treatment (if any) applied in manufacturing the food product,

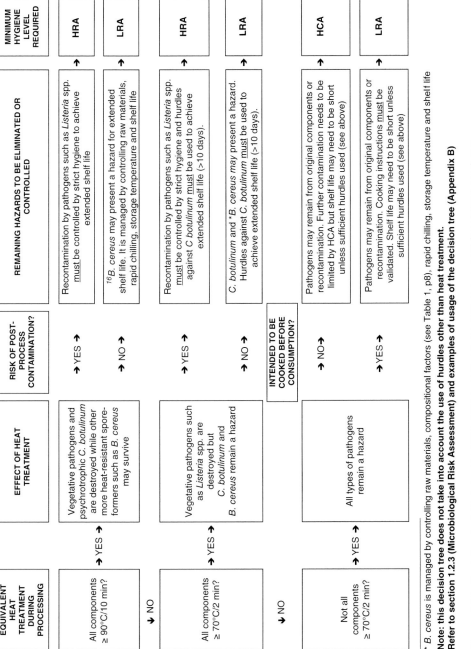

* *B. cereus* is managed by controlling raw materials, compositional factors (see Table 1, p8), rapid chilling, storage temperature and shelf life

Note: this decision tree does not take into account the use of hurdles other than heat treatment.

Refer to section 1.2.3 (Microbiological Risk Assessment) and examples of usage of the decision tree (Appendix B)

Figure 44.1 Decision tree to determine the minimum hygienic status required for chilled products.

any microbiological hazards arising between the heat treatment and final pack sealing and the final status of the food (i.e. ready to eat, ready to reheat or ready to cook) must be considered.

In general, only cooked-in-pack products are free of any risk of recontamination between heat treatment and final pack sealing and can therefore be produced in an LRA. The completed pathway through the decision tree shows the **minimum** hygiene status of the product handling environment between heat treatment and final pack sealing.

44.19   A detailed description of the requirements in each of the above areas is given in the CFA's *Best Practice Guidelines for the Production of Chilled Food* (4th edition, 2006). The standards highlighted are **minima** for each type of product. Higher standards can be used, but due regard should be paid to the potential for cross-contamination between lower-status and higher-status products within a single-status environment. Other products must not be produced in an HCA unless HACCP shows there are no additional risks to all products. Expert microbiological advice should be sought if necessary.

Quality Control   44.20   Microbiological testing of raw materials, intermediate products and finished products should aim at monitoring and verifying standards and trends, not simply accepting or rejecting on the basis of results. In cases of unsatisfactory results, recall procedures would be necessary; therefore any adverse trend in results calls for immediate investigation of process conditions and controls, including in the supply chain, thus reducing the risk of a recall situation developing (see Chapter 27). It should be determined whether a positive or negative release system is in place with regard to microbiological testing.

44.21   Production programmes should provide sufficient time for adequate cooling; a deadline for acceptance of late orders should be established to ensure this. Care should be taken to avoid condensation on cold product by exposure to warm product or humid conditions.

44.22   The correct functioning of refrigerated despatch vehicles should be checked before loading, and product temperatures should be checked and marked on delivery notes before loading. At maximum load, despatch vehicles should be able to maintain specified temperatures and ensure that products carried remain within appropriate temperature profiles throughout the load and for the whole journey. A system should be in place to monitor the vehicle's operating temperature. This could include manual logging of data on the appropriate record during transport of the product, inspections using a calibrated temperature probe at each delivery point (where the temperature is formally recorded) or data logging devices where the data can be electronically downloaded.

All equipment must be checked at routine intervals to verify that the results obtained are accurate and the equipment is working within an acceptable tolerance (see Chapter 34). Documented transport breakdown procedures should be in place that define the actions to be taken in the event of vehicle or refrigeration unit breakdown, and the records to be maintained that detail the incident and corrective action taken.

44.23 The quality control function should liaise with distribution management and customers to establish that there is adequate provision for maintenance of the temperature-controlled chain during transport and storage with the need for this being understood by all personnel. Major control points also include the occasions when products may not be in a temperature-controlled environment during the loading and unloading of vehicles, chilled stores and retail display cabinets.

44.24 Any frustrated deliveries that are returned should be examined by quality control, who should ensure that only goods in standard condition are accepted for further despatch. Other returns, suitably labelled, should be kept separately until disposal is agreed and actioned.

High-risk Areas 44.25 Special conditions must be provided for the production of high-risk products. Detailed requirements are set out in the CFA Best Practice Guidelines. The production area should be completely separated from other areas from floor to ceiling. Positive pressure ventilation with micro-filtered air at the appropriate temperature and humidity should be provided. However, rather than a specific pressure being recommended, which is technically challenging to measure, a minimum number of air changes per hour is the preferred control parameter. Entry and exit should be through changing rooms, with 'no-touch' washing facilities. Emergency exits from the production area should be alarmed so that access via these exits can be monitored. Personnel must wear, as instructed, designated internal footwear and clothing. Only fully processed food components and packaging materials should be admitted, with entry through hatches/airlocks. Chill rooms for products awaiting prepacking should be part of the designated special area. Construction and finish should provide for easy cleaning and disinfection. Only essential equipment and materials should be permitted in the production area. External packaging and ingredients entering the area may be subject to a sanitising procedure. Production should be interrupted for frequent thorough cleaning and disinfection. The frequency should be determined by risk assessment, which should be formally recorded and retained together with any results of verification activities such as product sampling or swab results. Otherwise cleaning and disinfection or sanitation may be undertaken between product types or varieties in addition to other hygiene activities, for example between the assembly of different sandwich types.

44.26 Only named, authorised and medically cleared individuals should enter the production area. This includes management, technicians, engineers and visitors. It must be impressed on visitors, contractors and staff how critical it is to maintain personal hygiene and minimise the risk of product contamination (see Chapter 17). Medical screening should be undertaken with emphasis on:

(a) skin health – absence of acne, boils, other infections, wounds and burns;
(b) ear, nose and throat infections;
(c) gastrointestinal disorders; and
(d) contact with known cases of food poisoning.

44.27 Pre-employment screening should be made by a physician with experience in occupational medicine. The health status of selected employees should be reviewed by the physician at appropriate intervals, and a daily check, before starting work, should be made by an occupational nurse or designated individual.

44.28 Before entering the special area, outer garments should be changed for clean overalls and internal footwear. Effective head covering, coloured distinctively, should be worn. Beards and long hair should be discouraged, but if worn should be fully contained. Hands should be washed with a non-aromatic antibacterial soap or lotion. Personnel should not use the same catering or toilet facilities as personnel from other areas (see Chapter 17).

44.29 Designated equipment should be provided in each area to prevent cross-contamination. This includes maintenance equipment that should be designated to specific areas, for example HRA, so that the tools can only be used in that area.

45   FROZEN FOODS

Principle

Frozen foods are preserved by freezing and storing at temperatures cold enough to inhibit the growth of micro-organisms and to retard chemical and physical reactions to a negligible rate. Quick-frozen foodstuffs can be described as those foods that are subjected to the 'quick-freezing' process, in which the temperature zone of maximum crystallisation is spanned as rapidly as possible and the product is then held (after thermal stabilisation) at a temperature of –18 °C or lower.[1]

General

45.1   Hazard analysis critical control point (HACCP) should be used as a management tool to develop a HACCP plan and associated food safety management system (FSMS) and quality management system (QMS). At critical control points (CCPs) and critical quality points (CQPs), critical limits and target levels and tolerances should be established so that a tendency towards a loss of control can be detected and rectified rather than waiting until a potentially unsafe product has been produced (see 45.12).

Personnel and Testing   45.2
Facilities

Personnel engaged in management, production, quality control and maintenance should receive the necessary training and conform to the medical and hygiene requirements described in Chapter 17. Production should be supervised by appropriately qualified and skilled personnel. Facilities should be available for microbiological and quality testing.

Raw Materials

45.3   Raw material specifications, acceptance and storage should be in accordance with the provisions stated in this Guide.

45.4   Raw materials should be selected for their ability to withstand freezing and thawing cycles, for example using specially modified starches to prevent syneresis.

45.5   Incoming high-risk foods such as meat, poultry and fish should be subjected to microbiological testing. Such raw material should be temporarily quarantined until test results are available, that is, a positive release system must be in place.

45.6   Although microbial load may sometimes be reduced by freezing, this cannot justify the use of raw materials that have a high microbiological loading. Freezing does not destroy any microbial toxins that may have been formed, and does not eliminate the possibility of microbial problems at a later stage. Enzymes that have not been inactivated by treatment before freezing may continue to act after freezing with the associated reduction in product quality.

[1] http://eur-lex.europa.eu/legal-content/EN/TXT/?uri=LEGISSUM:l21116.

Food & Drink – Good Manufacturing Practice: A Guide to its Responsible Management, Seventh Edition.
The Institute of Food Science & Technology Trust Fund.
© 2018 John Wiley & Sons Ltd. Published 2018 by John Wiley & Sons Ltd.

45.7 The freezing of food will not improve its quality. The flavours of some spices and other flavours can change quite drastically on freezing and subsequent storage. In products where enzyme systems are not fully inactivated, freeze concentration of enzymes and other substances can cause damage to texture and flavour. The sensory evaluation step of product development should be done after passing through one freeze/thaw cycle, with ultimate product life requirements being taken into account.

45.8 Preparation of raw material should be carried out in areas segregated from those concerned with cooked, blanched or finished product. Associated training programmes must underpin why such segregation is important for product safety and quality.

45.9 High-risk raw material, for example poultry, meat or fish, should be prepared in rooms completely isolated from cooked or finished products by walls from floor to ceiling. There should be no interchange of staff or equipment between the areas. The preparation rooms should be maintained at a cool temperature and provided with air, filtered to remove micro-organisms, at 8 °C under positive pressure.

45.10 For the production area and equipment, a 'clean-as-you-go' regime should be instituted, debris and gross soiling being constantly removed, and each day a major strip-down and cleaning and disinfection operation should be carried out according to a validated, documented cleaning schedule.

45.11 The only cryogenic media that should be used are air, nitrogen and carbon dioxide that meet purity criteria set by the European Commission.

45.12 CCPs should be established at points in the process where microbial proliferation could occur if suitable controls are not in place. All critical limits and target levels and tolerances at CCPs should be validated. Monitoring and verification programmes for these CCPs should be established and implemented by trained staff with the knowledge and skills required to ensure such actions are effective. Monitoring activities are those measurements and tests that are undertaken whereby if non-conforming a process, material and/or product can be brought back under control within the production cycle, i.e. before it has left the manufacturing unit. All other forms of testing and measurement of a material, product or process are verification activities. Monitoring and verification of CCPs could involve:

(a) measurement of the temperature and residence time of a particular material at frequent intervals throughout the shift;
(b) assessment of the degree of build-up of food debris or holding time of food material prior to or during processing, at frequent intervals throughout the shift;

(c) collection and microbiological testing of swabs or food samples at designated intervals during the production run. If the product is positively released on the basis of the sample results, this is monitoring; otherwise if a negative release system is in place, this is a verification activity.

(d) assessment to determine the presence of physical hazards in the product, for example metal detection, magnets, x-ray or imaging equipment.

It is critical that as a result of an effective training programme production staff understand whether the assessment activities they are undertaking are monitoring or verification and the corrective action required in the instance of non-conformance. Corrective and preventive action may involve holding and quarantining product on the manufacturing site as a result of a problem being identified via monitoring or the initiation of product withdrawal or recall in the event that a failure is identified as a result of verification activity. Timeliness of decision making is fundamental to effective good manufacturing practice (GMP) being demonstrated.

Results of monitoring and verification should be recorded, preferably as graphs or histograms, when considering process or product attributes, and then assessed for signs of trends. These results should form a basis for management review and internal audit programmes (see Chapter 11) so that preventive action, and if required corrective action, can be implemented to address potential weaknesses in the FSMS and QMS (see Chapter 28).

Hygiene audits should also be analysed for trends, especially if this includes further verification by swab analysis. From such records, the key performance indicator (KPI) or critical success factor (CSF) for a particular line can be established, and any adverse departure from these indicators must be countered by implementing corrective action to improve hygiene. It should be borne in mind that if the product has to be thawed before cooking by the end user, high mould and yeast levels within the food can be a special problem.

45.13 Products or ingredients that are heated or cooked during manufacture should be cooled as quickly as possible to below 8 °C. Throughout the preparation stage, every effort should be made to keep the food cool and to keep it progressing along the production line.

Quality Plan

45.14 A quality plan should also be implemented to monitor raw material and product quality attributes at quality points and CQPs in the process with the associated monitoring and verification activities as identified above.

Storage

45.15 Storage conditions and controls are addressed in Chapter 26. Frozen storage must be adequately controlled to prevent excessive build-up of ice on walls, floors and ceilings. Consideration should be given to personnel health and safety when working in these areas, including the provision of alarms if personnel can be locked inside the store.

45.16 Primary long-term (3 months or more) stores should be maintained at or colder than −26 °C. Temperature fluctuations of more than ±2 °C should be avoided and the frequency of the variation kept to a minimum. The monitoring and continuous recording of storage air temperature using calibrated probes is essential. The number and location of temperature probes should be such as to ensure effective monitoring. Air temperatures in cold stores should also be manually checked and logged at least once every 24 hours. If conditions warmer than specified warning and action limits are identified, then the prescribed corrective action must be followed. Records should be maintained that detail any incident and the corrective action taken. Raw material and product temperatures should also be monitored to ensure that they remain within acceptable levels. Records should be retained of such activities. The temperature of raw materials and products should not be higher than −18 °C. If the products have a temperature above −18 °C, they should be quarantined, inspected, records maintained and appropriate action taken.

Freezing

45.17 Rapid and well-controlled freezing without delay after preparation is essential, at a rate appropriate to the specific product. Mechanical damage to the texture of cellular products can result from slow growth of large ice crystals.

45.18 Freezing equipment should be designed to give the minimum amount of product dehydration in order to prevent dehydration damage on the outer surface of the product.

45.19 Freezing should not be regarded as complete until a temperature of −18 °C has been reached at the thermal centre of the food. The degree of homogeneity of the food needs to be considered, especially if it is of a particulate nature of varying sizes. Freezer exit temperature should be such that the temperature of the products should not rise above −18 °C during subsequent packing and palletisation. Some products can suffer surface cracking with too rapid freezing, for example pastry products, and in this case freezing will be a compromise of the process to give maximum safety control and quality to the centre phase while keeping the outer phase intact.

Packaging

45.20 Food packaging should comply with the general requirements stated in Chapter 24, and also include a water vapour barrier to prevent dehydration and weight loss during storage.

45.21 Packaged frozen foods must carry the storage instructions necessary to validate the stated indication of minimum durability that has been determined by the manufacturer.

45.22   Precautions should be taken to ensure that packing integrity is maintained during processing, storage or distribution to avoid risk of product contamination, product rejection or freezer burn, which is caused by product dehydration.

45.23   When packing in bulk for later repacking, sacks or palletainers should be labelled with production date, process details, and variety and production batch to facilitate identification at a later date. End-user packs should be labelled with repack dates (coded if necessary) and duration dates. A record should be kept of materials that have been sent to individual repacking points so that end-user packs can be traced back to the actual production records if required (see Chapter 14).

45.24   Primary long-term (3 months or more) stores should be maintained at or colder than −26 °C. Temperature fluctuations of more than 2 °C should be avoided and the frequency of variation kept to a minimum. The monitoring and recording of storage unit air temperature using calibrated temperature probes is essential. The number and location of temperature probes should be such as to ensure effective monitoring. Air temperatures in cold stores should be manually checked by quality control personnel and logged at least once every 24 hours. Temperature warning and action limits should be specified and responsibilities for taking corrective action defined. Records should be maintained that detail any incident and the corrective action taken.

45.25   During delivery to the primary cold stores, products should not be exposed to direct sunlight, wind or rain; transfers should be carried out with the minimum of exposure to outside temperature conditions. Products incoming to the stores should not be warmer than −15 °C and should remain in the stores until their temperature has reached −23 °C or colder. Products leaving the stores should be protected as before from adverse weather or temperature conditions; as much as possible of the pre-removal operation should be carried out in the cold store. To afford maximum protection to the products during unloading and loading refrigerated transport, it is advisable to fit the stores with a loading bay with suitable protection positioned so that the temperature-controlled vehicle can be brought into direct contact with the store. The same principles should apply to retailers and caterers where there is a cold receipt area.

45.26   Secondary or distribution cold stores should be maintained at −23 °C or colder. Temperature probes that monitor storage temperatures should be so positioned as to ensure effective monitoring. To this end, it is important that the relationships between core and surface temperatures are understood and their implications within the distribution and storage system taken into account. An alarm system should be operational to warn of any breakdowns. All autographic recorders and temperature probes should be accurate to within 1 °C and ideally 0.5 °C, and be calibrated at predefined intervals in such a way as to give traceability back to a national standard. If an autographic system is not available, manual logs should be maintained.

45.27 Cyclic variation of temperature, whether above or below the recommended temperature, is undesirable as it accelerates dehydration even in hermetically sealed packs when it causes migration of moisture from product to form 'snow' inside the packaging.

45.28 Clearly defined procedures should be laid down to deal with wide temperature fluctuations or refrigeration breakdown. Should the air temperature rise to −12 °C, the cold store should be closed until the fault has been cleared and the correct store temperature restored. Products in the store during the period of breakdown should be checked by the quality control manager or designate and records maintained of the decision for product disposition and associated actions, including the removal of product to an alternative store.

45.29 Primary distribution vehicles should be designed so as to maintain the same temperature as the primary cold stores, and operated in such a way that the delivery temperature of the product is not higher than −18 °C. If the products have a delivery temperature above −15 °C, their quality should be checked, records maintained and appropriate action taken. Documented transport breakdown procedures should be in place that define the actions to be taken in the event of vehicle or refrigeration unit breakdown, and the records to be maintained that detail the incident and the corrective action taken.

45.30 Secondary distribution vehicles are those used for transporting products between secondary cold stores and point of sale. They should be designed and operated in such a way that the food can be delivered at −15 °C. If the product temperature is above −12 °C, the overall quality should be checked, records maintained and appropriate action taken. Documented transport breakdown procedures should be in place that define the actions to be taken in the event of vehicle or refrigeration unit breakdown, and the records to be maintained that detail the incident and the corrective action taken (see 32.17).

45.31 Temperature monitoring and recording in all distribution vehicles and holding centres should be such that relevant personnel are able to verify the temperatures of vehicles and cold stores. Further recommendations are therein given for the transport, handling, storage and display by the wholesaler and retailer, and it is to the producer's ultimate advantage to encourage the observation of such recommendations (see Chapter 32).

Legislation   45.32 Reference may also be made to the British Frozen Food Federation website for further guidance (http://www.bfff.co.uk)

and to the **Ice Cream (Heat Treatment, etc.) Regulations 1959, SI 1959 No. 734 as amended by SI 1962 No. 1287** and **SI 1963 No. 1083**, to the **Quick-frozen Foodstuffs Regulations 1990, SI 1990 No. 8136, as amended by SI 1994 No. 298** and the **Quick Frozen Foodstuffs Regulations 2007**, that are implemented in the UK, and the EU Directives on the approximation of the laws of the Member States relating to quick-frozen foodstuffs for human consumption (**89/108/EEC, 92/1EEC** and **92/2/EEC as subsequently amended**). Associated EC legislation includes **(EC) No. 37/2005 of 12 January 2005** on the monitoring of temperatures in the means of transport, warehousing and storage of quick-frozen foodstuffs intended for human consumption.

45.33 The **Quick Frozen Foodstuffs Regulations 2007 (implementing EC 37/2005)** and replacing the **Quick Frozen Foodstuffs Regulations 1990** prohibit the placing on the market of a quick-frozen foodstuff unless certain conditions are satisfied. Specific legislation for quick frozen foodstuffs (QFFs) does not apply to ice-cream and edible ices. The regulations require that all temperature monitoring instruments used in the transport, warehousing and storage of QFFs must comply with European standards (**EN12830, EN13485** and **EN13486**). All relevant organisations must keep documentation demonstrating that instruments meet relevant European standard(s). QFF produced by a manufacturer that is to be marketed in its packaging to the ultimate consumer or to a caterer must include on the label, in addition to normal labelling requirements:

- 'quick frozen';
- the date of minimum durability – a 'best before' date;
- the maximum advisable storage time;
- the temperature or the equipment that should be used to store it;
- a batch or lot mark; and
- wording such as 'do not refreeze after defrosting'.

QFF packed by manufacturers for further processing must include on the label:

- 'quick frozen';
- a batch or lot mark; and
- the name (or business name) and address of the manufacturer or packer, or a seller in the EU.

45.34 QFFs must be prepacked in a way that prevents contamination, deterioration and dehydration of the food material. Temperature records must be traceable to a given date, and by inference the batches of food stored and these records must be retained and be retrievable for at least one year or longer, depending on the nature and shelf life of the QFFs they relate to.

46   DRY FOOD PRODUCTS AND MATERIALS

Principle

Dry food products or materials are those foods that have a low moisture content. Some materials are naturally dry. Alternatively, the removal of water from a food material is undertaken during manufacturing in order to inhibit the growth of bacteria, yeasts and moulds. Examples of manufacturing processes that will dry foods include desiccation, dehydration and evaporation processes such as freeze drying, spray drying, air drying and smoking. Dry goods production involves the same general hygiene considerations as other foods with regard to statutory requirements. These have been discussed in preceding chapters.

Microbiology

46.1   Most bacteria require a water activity (a_w) of at least 0.95, and very few micro-organisms can grow in foods around 0.6. Some bacteria, for example *Bacillus cereus*, can produce spores that will germinate on the reconstitution of the dry product. Yeasts and moulds can grow at lower levels of a_w.

46.2   Dry foods and their ingredients can carry heavy microbiological loads undetected and without the dry food undergoing any noticeable change. This can be exacerbated by incubative temperatures, which can occur during some drying and mixing operations. Vegetative organisms or spores can proliferate and cause problems when the dry food is reconstituted. Dried materials used as the ingredients in the manufacture of canned products should have low bacterial spore counts. All ingredients, according to their origin and substance, should be subjected to a system of microbiological control by examination. Similarly, finished products should be sampled within a scheme of microbiological examination. When the food is to be consumed without further heat processing, it must be microbiologically safe. Particular care should be taken to prevent airborne dust from causing microbiological contamination of raw materials, finished product and plant. In areas that are dry-cleaned, microbiological contamination can be very difficult to remove.

General Hygiene

46.3   Processes involving dry materials can have problems associated with dust, particularly those of cleaning, the possible creation of an explosive dusty atmosphere and the risks of cross-contamination by dust particles. It is important therefore to contain dust as far as possible in an enclosed system and, with the aid of dust removal and extraction systems, to maintain a high standard of cleanliness. If allergens are utilised as raw materials, then dust control is critical and control measures need to be in place according to the physical state of the material, including particulate size and hardness of the particle.

Food & Drink – Good Manufacturing Practice: A Guide to its Responsible Management, Seventh Edition.
The Institute of Food Science & Technology Trust Fund.
© 2018 John Wiley & Sons Ltd. Published 2018 by John Wiley & Sons Ltd.

46.4   The general environment of the plant and equipment, including ledges and girders, should be regularly cleaned and an effective air extraction system should be installed. Such a system should discharge through a filter and at a point situated so as to minimise the risks of the discharge being able to contaminate the other plant or products. Dust extraction systems should be properly maintained, cleaned and serviced; they become heavily coated inside duct-work, and cleaning and filter changing can create a very dusty atmosphere. Dusty atmospheres should be considered as potentially dangerous explosion hazards. It may be desirable therefore to use flameproof motors and switches or ensure that they are situated in a relatively clean environment. Adequate protective clothing and other equipment should be provided for those involved physically in the cleaning operations, and during production if necessary.

46.5   Plant and equipment should be designed with easy access for cleaning in mind. When possible, manufacturing operations should be carried out in closed vessels or systems. Spray and fluid-bed dryers should be fitted with efficient filter bags. Where closed systems are not practicable, it is usually possible to carry out an operation within a dust extraction system. Delicacy of handling in relation to product friability may also need special attention.

46.6   Particle size reduction of dry materials produces fine dust, and unless the process can be carried out in a closed system, including the discharge of the ground product, it should be operated within a dust extraction system. In flour mills where stone milling is undertaken, controls should be in place for the assessment of foreign body contamination risk. Further consideration should be given to the level of contamination present in wheat and the use of in-line magnets and screens to remove foreign bodies. Care must be taken to ensure that the product does not 'bridge' at these screening points. An investigation should be implemented where necessary to identify the likely source of the contaminant so that appropriate corrective action can be identified and instigated. Mixing and sieving operations usually have similar dust control problems, and these tasks should be undertaken using appropriate dust control methods.

46.7   A particular problem occurs with hygroscopic materials that become sticky and so call for special attention with regard to cleaning. They are not always completely removed by dust extraction systems, but those particles that are removed can clog ducts and filters. If a vacuum system is used for dry cleaning, then care must be taken to ensure that all pipework and attachments are stored in suitable locations in each production area and routinely inspected for signs of wear and damage and replaced as necessary. This inspection should be formally recorded. Wet cleaning methods should be available in all areas where such dry goods are produced and handled, but precautions should be taken against the risk of creating humid conditions, which might allow microbial growth.

Contamination 46.8 Dry materials require special attention concerning contamination and its detection. Particular attention should be given to examination for insect infestation, both past and present, and measures adopted to deal with either or both situations. At all times dry materials should be protected against attack by insects and rodents. Dry goods should be inspected on arrival for signs of pest infestation, including stored product insect contamination. There have been national recalls in the UK, for example, for both weevil and biscuit beetle contamination, but other pests such as psocids (booklice), rust red flour beetle, flour moths and their larvae are also of concern. Visual inspection will include observing the materials, delivery vehicles and subsequently storage areas for signs of pests or infestation. Where materials/ premises are found to be infested, the pest contractor must be informed and appropriate action taken. Any contaminated material must be disposed of immediately to prevent the further spread of contamination.

While the use of magnets has been previously discussed, all dry materials should pass through a metal detector at least once.

Packaging 46.9 Consideration must be given to the potential minimum durability
and Storage of dry products in the form in which they are offered for sale. Exposure of such a product to light, air and water vapour may cause physical or chemical or both types of change in the product, sometimes fairly rapidly, as may significant changes of temperature. Packaging can protect the product from these effects and can also protect it from insects and rodents; it should therefore be considered and designed with these objectives in mind.

46.10 Where dry foods are intended to be reconstituted for consumption, detailed instructions for preparation given on the package should, where applicable, have regard to safety considerations.

47　COMPOSITIONALLY PRESERVED FOODS

Principle

Some foods, such as jams and pickled products, depend for their preservation and/or specific properties and maintenance of their quality during their expected life on achieving a particular quantitative composition (e.g. the attainment of requisite – and, in the UK and the other European Union (EU) Member States, also legally required – refractometric solids in jams with no added preservative, or of a preservation index of 3.6% acetic acid in the volatile constituents in non-pasteurised pickles and sauces).

47.1　In products where a quantitative compositional factor is critical, the training of production supervisors, operators and quality control staff should emphasise the critical nature of such compositional factors. Production methods and control procedures should be such as to ensure that the required composition is consistently achieved. Any production batch that is found to be non-conforming should be quarantined and dealt with in accordance with the procedures outlined in Chapters 28 and 29.

47.2　Where appropriate, relevant on-pack instructions for use should include instructions for storage.

Food & Drink – Good Manufacturing Practice: A Guide to its Responsible Management, Seventh Edition.
The Institute of Food Science & Technology Trust Fund.
© 2018 John Wiley & Sons Ltd. Published 2018 by John Wiley & Sons Ltd.

48   FOODS CRITICALLY DEPENDENT ON SPECIFIC INGREDIENTS

Principle

Some foods are critically dependent for their preservation and/or specific properties and maintenance of quality during their expected life on the presence in the required amount(s) of one or more particular specific ingredients, usually at relatively low levels.

General

48.1   All product specifications should be validated as part of the design process before full production commences. Revalidation is required if there is a change to product formulation, ingredients supplier and so on to ensure that food safety will not be compromised (see Chapter 3). All ingredients must be declared 'food grade', and there must be documented evidence held by the manufacturer to substantiate this. Quantitative Ingredient Declarations (QUID) labelling must be complied with at all times for prepacked foods.

48.2   Where the preservation and/or specific properties of the product critically depend on the inclusion, at a specified level, of one or more specific ingredients, the importance and critical nature thereof should be emphasised in the training of production supervisors, operatives and quality control staff. Production methods and control procedures should include specific provision directed towards ensuring that the substances concerned are not accidentally omitted, nor added in incorrect quantity, nor unevenly distributed through the product, whether resulting from operator error or inaccurate functioning of automatic dispensing equipment. Whereas a deficiency would impair the preservation and/or specific properties of the food, an excess might, in the case of certain specific additives, exceed specified legal limits and/or be harmful to the consumer.

48.3   Because of the wide variety of possible combinations of circumstances, it is impossible to generalise as to specific production techniques or control measures required for the purpose indicated in 48.2. Where discrete batch quantities of any specific ingredient are required, it may be appropriate for these to be accurately weighed by or under direct supervision of laboratory or quality control staff. Where colour or a coloured ingredient is also part of the product formulation, consideration should be given to the feasibility of premixing or adding in the specific ingredient therewith, so that the visually obvious presence or absence of the colour in the batch acts as a 'marker' for the specific ingredients. Batch manufacturing records (see Chapter 13) should include provision for recording the addition of such substances, and operator training should emphasise the importance of filling

Food & Drink – Good Manufacturing Practice: A Guide to its Responsible Management, Seventh Edition.
The Institute of Food Science & Technology Trust Fund.
© 2018 John Wiley & Sons Ltd. Published 2018 by John Wiley & Sons Ltd.

these in at the time of making the addition and not filling in a succession of ticks and initials at some later convenient time. It is very important that the operator initials the entry for each batch so that in the event of non-compliance there is traceability to a particular operator (see 13.8). At the end of a production run, there should be a cross-check of the usage of the critical specific ingredients against the amount(s) required by the actual production volume during the run. Any discrepancy between the actual and expected amounts of a specific ingredient should be investigated.

Generally speaking, the natures of the substances likely to be involved do not lend themselves to continuous or rapid monitoring, so that analytical control provides only intermittent spot checks, which should, however, be carried out to confirm the effectiveness of the manufacturing disciplines on a retrospective basis. It is important to check the identity of raw material supplies, that is, that they are in fact the substances that they purport to be. It is evident that this assumes additional importance in the case of the specific ingredients referred to above.

In the event of identification of non-conformance then corrective action procedures must be implemented (see Chapter 28) and if required product withdrawal or product recall instigated (see Chapter 27).

49   IRRADIATED FOODS

Principle

Food irradiation is the controlled process that is a physical treatment of food with high-energy ionising radiation. The use of irradiation is limited in many countries. If irradiation is used, it must be an authorised activity in the country where it takes place. Irradiation is a process by the manufacturer that must be identified on the packaging and documentation accompanying the designated batch in line with legislative requirements in the country where the operation takes place and the countries to which the product may be exported.

General

49.1   Irradiation must present no health risk and must not be used as an alternative to good sanitary conditions in the food supply chain, or to replace the elements of good manufacturing practice (GMP) defined in this Guide. Irradiation may be used to prevent germination and sprouting of vegetables such as onions, garlic or potatoes, to slow down the ripening, ageing and senescence of fruit and vegetables, and to destroy pathogenic or spoilage bacteria, moulds, viruses or insects and by doing so prolong the shelf life of food and minimise the likelihood of foodborne illness and food poisoning. The foods and food ingredients authorised for irradiation in the European Union (EU) are identified in a given list (see https://ec.europa.eu/food/safety/biosafety/irradiation/legislation_en). The national authorisations that are deemed valid are identified by the EU as well as the standardised analytical methods for the detection of irradiated foods which have also been adopted by Codex Alimentarius.[1] The EU List of Approved Premises for Irradiation was last updated in 2015.[2]

49.2   **Directive 1999/2/EC – Irradiation of Food and Food Ingredients** lays down general provisions and the rules governing approval and control of irradiation and changes the rules on the labelling of foodstuffs that have been treated with ionising radiation. **Directive 1999/3/EC** establishes an initial positive list of foodstuffs that can be treated and freely traded within the EC. **Document 2002/840/EC** identifies the list of approved facilities in third countries for the irradiation of food as amended by **2004/691/EC, 2007/802/EC** and **2010/172/EC** and **Commission Implementing Decision 2012/277/EU**.

49.3   In the United Kingdom (UK) The **Food Irradiation (England) Regulations 2009**[3] or its equivalent implementing **EU Directive 1999/2/EC** and **1999/3/EC**, as amended by the **Commission Decisions** above, contains mandatory labelling requirements on

---

[1] https://ec.europa.eu/food/safety/biosafety/irradiation/legislation_en.
[2] http://eur-lex.europa.eu/legal-content/EN/ALL/?uri=CELEX:52015XC0213(02).
[3] http://www.legislation.gov.uk/uksi/2009/1584/pdfs/uksi_20091584_en.pdf.

Food & Drink – Good Manufacturing Practice: A Guide to its Responsible Management, Seventh Edition.
The Institute of Food Science & Technology Trust Fund.
© 2018 John Wiley & Sons Ltd. Published 2018 by John Wiley & Sons Ltd.

irradiated food sold. The **Food Irradiation (England) (Amendment) Regulations 2010** made further amendments to the legislation and updated the list of approved food irradiation facilities.[4] The overall average dose calculation is defined in the **Food Irradiation (England) Regulations 2009**, but it should be noted that while the dose can be readily determined for homogenous products, it is more difficult in a heterogeneous product where particulate size or density varies. In some cases, the mean value of the average values of the minimum dose measured in the product (D_{min}) and the maximum dose (D_{max}) will be a good estimate of the overall average dose. In this instance, the overall average dose is ($D_{max} + D_{min}$)/2. The ratio of D_{max}/D_{min} (dose uniformity ratio) must not exceed 3 and the maximum absorbed dose must not exceed 150% of the overall average dose.

49.4 The following types of ionising radiation are allowed by UK legislation:

(a) gamma rays from the radionuclides cobalt-60 or caesium-137;
(b) X-rays generated from machine sources at or below an energy level of 5 million electron-volts (MeV); and
(c) electrons generated from machine sources at or below an energy level of 10 MeV.

49.5 The food is passed through a field of ionising energy and the ionising radiation passes through the food, generating large numbers of short-lived free radicals. These can kill micro-organisms, such as Salmonella, and inhibit the processes described in 49.1. At no time during the irradiation process does the food come into contact with the radiation source and, by using cobalt-60 or electron beams up to 10 MeV, it is not possible to induce radioactivity in the food. The length of time the food is exposed to the ionising energy and the strength of the source determine the irradiation dose, measured in kilograys (kGy), the food receives.

49.6 In the UK, only foods that fall within the following classes can be considered for irradiation:

Maximum overall average dose in kilograys (kGy)[5]

Fruit (including fungi, tomatoes, rhubarb)	2
Vegetables (including pulses)	1
Cereals	1
Bulbs and tubers (potatoes, yams, onions, shallots, garlic)	0.2
Dried aromatic herbs, spices and vegetable seasonings	10
Fish and shellfish	3
Poultry	7

---

[4] https://www.food.gov.uk/business-industry/imports/importers/irradiated.
[5] https://www.food.gov.uk/science/irradfoodqa.

These categories of food can first be irradiated and then used as ingredients in other food products. A mixture of foods from the same class, for example a blend of herbs and spices, can be irradiated. A composite food product is not permitted for irradiation treatment unless minor ingredients (including additives) together comprise no more than 2% by weight, excluding any added water from the calculation.

Supplier approval of ingredients should pay particular attention to whether spices have been irradiated at source not only to ensure appropriate labelling of the finished product but also as part of raw material risk assessment. *Salmonella* contamination of foods has been linked to ingredients such as ground black pepper.

Facilities and Control 49.7
of the Process

Irradiation treatment of foods must be carried out in facilities licensed and registered for this purpose by the appropriate authority. Irradiation procedures must be fully validated and records retained to demonstrate their efficacy and their compliance with legislation. Revalidation is required whenever the product, its geometry or the irradiation conditions are changed. This revalidation process must be formally recorded.

During the process, routine dose measurements are carried out in order to ensure that the dose limits are not exceeded. Measurements should be carried out by placing dosimeters at the positions of the maximum or minimum dose, or at a reference position. The dose at the reference position must be quantitatively linked to the maximum and minimum dose. The reference position should be located at a convenient point in or on the product, where dose variations are low.

Routine dose measurements must be carried out on each batch and at regular intervals during production. In cases where flowing, non-packaged goods are irradiated, the locations of the minimum and maximum doses cannot be determined. In such a case, it is preferable to use random dosimeter sampling to ascertain the values of these dose extremes. Dose measurements should be carried out by using recognised dosimetry systems, and the measurements should be traceable to primary standards.

During irradiation, certain facility parameters must be controlled and continuously recorded. For radionuclide facilities, the parameters include product transport speed or time spent in the radiation zone and positive indication for the correct position of the source. For accelerator facilities, the parameters include product transport speed and energy level, electron current and scanner width of the facility.

49.8   The irradiation facility must be licensed for specific foods at specific doses (see 49.6).

49.9 A person seeking a licence to irradiate food should give the following details with the written application:

(a) the applicant's name;
(b) the applicant's address;
(c) the address of the facility at which the applicant proposes to irradiate food;
(d) details of any licence or registration under any other legislation that enables the applicant to use ionising radiation at the facility in circumstances where, but for that licence or registration, that use would be unlawful;
(e) a description of each food that the applicant proposes to irradiate which is sufficient to show that it falls within a permitted category of food;
(f) in respect of each food described pursuant to subparagraph (e):
   (i) the purpose for which the applicant proposes to irradiate the food and how that would benefit consumers;
   (ii) the method by which the applicant will ensure that the food is in a suitably wholesome state before irradiation;
   (iii) the overall average dose, maximum dose and minimum dose of ionising radiation that the applicant proposes to apply to the food;
   (iv) the method (including instrumentation and frequency) by which the applicant proposes to measure any dose of ionising radiation and the dosimetry standard that the applicant proposes to use to calibrate the dose meters used to measure it;
   (v) whether or not the applicant proposes to irradiate that description of food in packaging in contact with the food and, if so, the packaging that the applicant proposes to use; and
   (vi) whether or not the applicant proposes to apply temperature control to the food while irradiating it and, if so, the temperature at which the applicant proposes to keep the food during the application of temperature control.

49.10 The facilities must be designed to meet the requirements of safety, efficacy and good hygienic practices of food processing, in accordance with current statutory regulations in the UK and elsewhere and GMP. A list of licenced facilities is contained in Schedule 1 and Schedule 2 of the **Food Irradiation (England) (Amendment) Regulations 2010**.

49.11 Control of the process involves the food manufacturer and the operator of the irradiation facility (licence holder). Controls within the facility must include keeping records such as licence details, treatment and type of each food consignment, plus details of dose measurements. Additionally, each consignment must be accompanied throughout the processing chain by relevant documentation. Premises and records are open to inspection by the licensing authority. The licence holder is responsible for retaining records for a minimum of 5 years.

49.12 Incoming product must be kept physically separate from outgoing product.

Food Quality and Safety

49.13 The food before and after irradiation should comply with the provisions of the Codex Alimentarius Recommended International Code of Practice – General Principles of Food Hygiene and, where appropriate, with the Codex Recommended International Code of Hygienic Practice relative to the particular food.

49.14 Food intended for irradiation should be of a similar microbiological quality to that being processed by other means. For those foods irradiated to eliminate or reduce pathogenic organisms in the food, details of microbiological testing are required. **EU Regulation No. 2073/2005 on microbiological criteria for foods (as amended by EU Regulation No. 1441/2007)**[6,7] complements EU food hygiene legislation already described and applies to all food businesses involved in the production and handling of food. The microbiological criteria outlined can be used to validate manufacturing, handling and distribution processes and/or verify the acceptability of foodstuffs at specific stages in the process.

49.15 In the UK, public health requirements affecting microbiological safety and nutritional adequacy of specific foods must be observed.

49.16 Irradiation must not be used to process foodstuffs that do not meet the above microbiological quality standards.

49.17 Documentation referring to microbiological quality should accompany the food consignment from the manufacturer to the irradiation plant.

Application of the Process

49.18 Irradiation should only be applied to food manufactured in accordance with GMP, and with a demonstrable need for the particular use of the process in terms of food hygiene or other technological benefit. The doses applied should be commensurate with the application and within the set limit for the overall average dose for the food. Food manufacturers should define the correct dose to be applied, which should be adhered to by the irradiation plant operator.

49.19 Temperature control during distribution, handling and irradiation processing should be similar to the control exerted on analogous but non-irradiated foods. Irradiated food must be stored and transported with appropriate documentation that contains a statement to the effect that the food has been treated with ionising radiation.

[6] http://eur-lex.europa.eu/LexUriServ/LexUriServ.do?uri = CONSLEG:2005R2073:20071227:EN:PDF.

[7] http://eur-lex.europa.eu/LexUriServ/LexUriServ.do?uri = OJ:L:2007:322:0012:0029:EN:PDF.

49.20 Packaging materials should be of suitable quality and acceptable hygienic condition and be appropriate for the purpose. They should be able to withstand radiation in terms of their physical and sensory properties and in terms of migration of mobile components into the food. A full description of the materials is required. Packaging dimensions, particularly those of bulk packs, should take into consideration the nature of the irradiation source with respect to depth of penetration.

Re-irradiation

49.21 Food must not be re-irradiated. However, the full dose needed for a specific technological purpose may be given in sequential fractionated doses.

Import

49.22 Imports of irradiated foods into the UK are permitted only from countries that have been approved by the licensing authority. Lists of irradiation plants approved to supply specific irradiated foods and lists of countries approved to export specific irradiated food should be consulted. Appropriate documentation must accompany each consignment.

Export

49.23 Foods irradiated for export should comply with the guidelines described in Chapter 54.

49.24 Where the importing country's statutory requirements differ from the exporting country's standards, it is the responsibility of the suppliers to determine that their irradiated raw material or product will comply with these.

Labelling

49.25 In the UK, irradiated foods, whether prepackaged or not, must be labelled with the statement 'irradiated' or 'treated with ionising radiation'. This labelling requirement includes foods sold in catering establishments. Any irradiated ingredient incorporated into a food product necessitates the use of the same labelling terms in the ingredients list (see relevant food labelling regulations).

Principle

Novel food is defined by the European Union (EU) as food that has not been consumed to a significant degree by humans in the EU prior to 1997, when the first Regulation on novel food came into force. Further novel food can be described as newly developed, innovative food or food produced using new technologies and production processes as well as food traditionally eaten outside of the EU.[1] However, novel foods must be safe for consumption and labelled according to legislation in the countries in which they are produced and the countries to which they may be exported.

General

50.1   Care should be taken in the use of novel food or food ingredients produced from raw material that has not hitherto been used (or has been used only to a small extent) for human consumption in the area of the world in question, or that is produced by a new or extensively modified process not previously used in the production of food (and would thus include genetically modified materials). This must include attention to food safety considerations, compliance with the relevant regulations of the country for which the food is intended, and provision of label information to enable the purchaser or consumer to make an informed choice.

50.2   Novel foods and processes, including especially genetic modification (GM), have the potential to offer very significant improvements in the quantity, quality and acceptability of the world's food supply. Food scientists and technologists can support the responsible introduction of such foods and processes provided that issues of product safety, environmental concerns, information and ethics are satisfactorily addressed.

50.3   In the EU, the principal legislation is **EU Regulation No. 258/97** of the European Parliament and of the Council of 27 January 1997 concerning novel foods and novel food ingredients (OJ L43, 14.2.97, pp. 1–7). Proposals for amendment were tabled in 2008, 2011 and 2013, and agreement was reached on new legislation which was adopted on 25 November 2015.[2] The 2015 legislation states that 'food intended to be used for technological purposes and genetically modified food which is already covered by other Union acts should not fall within the scope of this Regulation. Therefore, genetically modified food falling within the scope of **Regulation (EC) No. 1829/2003 of the European Parliament and of the Council,**[3] food enzymes falling within the scope of

---

[1] https://ec.europa.eu/food/safety/novel_food_en.
[2] http://eur-lex.europa.eu/legal-content/EN/TXT/PDF/?uri=CELEX:32015R2283&from=EN.
[3] http://eur-lex.europa.eu/LexUriServ/LexUriServ.do?uri=OJ:L:2003:268:0001:0023:EN:PDF.

Food & Drink – Good Manufacturing Practice: A Guide to its Responsible Management, Seventh Edition.
The Institute of Food Science & Technology Trust Fund.
© 2018 John Wiley & Sons Ltd. Published 2018 by John Wiley & Sons Ltd.

Regulation (EC) No. 1332/2008 of the European Parliament and of the Council,[4] food used solely as additives falling within the scope of Regulation (EC) No. 1333/2008 of the European Parliament and of the Council,[5] food flavourings falling within the scope of Regulation (EC) No. 1334/2008 of the European Parliament and of the Council[6] and extraction solvents falling within the scope of Directive 2009/32/EC of the European Parliament and of the Council[7] should be excluded from the scope of this Regulation.' The legislation gives further definitions on nanomaterials, requirements for vitamins, minerals and other food supplements and food from animal clones which is regulated under Regulation (EC) No 258/97.[8]

50.4   Any release of genetically modified organisms (GMOs) into the environment is governed by national regulations implementing EU Directive 2001/18/EC [in the UK by the Genetically Modified Organisms (Deliberate Release) Regulations 2002], subject to amendments to the directive consequent on Regulation (EC) 1829/2003 on Genetically Modified (GM) Food and Feed, Regulation (EC) 1981/2006 and Regulation (EC) 298/2008. This has been further amended by Directive (EU) 2015/12 amending Directive 2001/18/EC, and Directive 2009/41/EC on contained use of genetically modified micro-organisms and Regulation (EC) 1946/2003 on transboundary movement of GMOs. For more information on current EU Legislation see https://ec. europa.eu/food/plant/gmo/legislation_en. The European Food Safety Authority (EFSA) is the central body controlling the assessment of novel foods and novel food ingredients. Applications can be made to a national authority [in the UK, the Advisory Committee on Novel Foods and Processes (ACNFP)]. Any person or company contemplating marketing in the UK a novel food or one containing a novel ingredient that has not already been the subject of official evaluation and approval must make a prior submission to the ACNFP.

50.5   Enforcement and execution of certain provisions of the EU Regulation and set fees for assessment procedures in the UK are provided by the Novel Foods and Novel Food Ingredients Regulations 1997 (SI 1997/1335) and the Novel Foods and Novel Food Ingredients (Fees) Regulations 1997 (SI 1997/1336). The Novel Foods and Novel Food Ingredients Regulations (1997) apply to the placing on the market within the EU of foods and food ingredients that have not previously been used for human consumption to a significant degree and that fall into the following categories:

[4] http://eur-lex.europa.eu/LexUriServ/LexUriServ.do?uri=OJ:L:2008:354:0007:0015:en:PDF.
[5] http://eur-lex.europa.eu/LexUriServ/LexUriServ.do?uri=OJ:L:2008:354:0016:0033:en:PDF.
[6] http://eur-lex.europa.eu/LexUriServ/LexUriServ.do?uri=OJ:L:2008:354:0034:0050:en:PDF.
[7] http://eur-lex.europa.eu/legal-content/EN/TXT/PDF/?uri=CELEX:32009L0032&from=EN.
[8] http://eur-lex.europa.eu/legal-content/EN/ALL/?uri=CELEX%3A31997R0258.

- foods and food ingredients containing or consisting of GMOs within the meaning of **Directive 90/220/EEC**, now superseded;
- foods and food ingredients produced from, but not containing, GMOs, foods and food ingredients with a new or intentionally modified primary molecular structure, now superseded as above;
- foods and food ingredients consisting of or isolated from microorganisms, fungi or algae;
- foods and food ingredients consisting of or isolated from plants and food ingredients isolated from animals, except for foods and food ingredients obtained by traditional propagating or breeding practices and having a history of safe food use; and
- foods and food ingredients to which has been applied a production process not currently used, where that process gives rise to significant changes in the composition or structure of the foods or food ingredients that affect their nutritional value, metabolism or level of undesirable substances.

The novel foods and food ingredients must not present a danger for the consumer, mislead the consumer or differ from the foods or food ingredients that they are intended to replace to such an extent that their normal consumption would be nutritionally disadvantageous for the consumer. Derogations are available for foods and food ingredients that, according to expert scientific opinion, are substantially equivalent to existing foods in respect of their composition, nutritional value, metabolism, intended use and level of undesirable substances contained therein. There are proposed **Novel Foods Regulations** planned to be implemented as legislation in 2018 that will repeal the **Novel Foods and Novel Food Ingredients (Fees) Regulations SI No. 1997/1336**. The proposed Regulations will enable the execution and enforcement in the UK of the **Novel Food Regulation (EU) 2015/2283** (which amends **Regulation (EU) No. 1169/2011** and repeals **Regulation (EC) No. 258/97** and **Commission Regulation (EC) No. 1852/2001**).

50.6 The EU legislation also specifies the assessment procedures that must be carried out before a novel food can be placed on the market and makes provision for objections to be raised by interested parties. The manufacturer should consult the relevant legislation in force at the time.

51 FOODS FOR CATERING AND VENDING OPERATIONS

<table>
<tr><td>Principle</td><td>The manufacture of foods intended for use in catering or vending operations should be carried out in accordance with the principles and practices outlined in this Guide but, additionally, should have regard to the special requirements that relate to the intended use. It should be noted that 'manufacture of food' in this context applies not only to food products made by a food manufacturing company and sold to a caterer or to a vending machine operator, but also to food prepared in a central production unit by factory-style processing, by a catering organisation for use in its own catering establishments, as distinct from preparation in 'cook-serve' form.</td></tr>
</table>

<table>
<tr><td>General</td><td>51.1</td><td>In the manufacture and distribution of food and drink for catering and vending purposes, particular regard should be given to the circumstances and conditions of use, the probable expertise of the caterer and his/her staff, and the interactions likely to occur between the product and its subsequent environment. The general points of guidance identified as pertinent to good manufacturing practice (GMP) apply equally to foods manufactured for catering and vending purposes. The manufacturer should be prepared to offer technical advice to users on the suitability of products for the uses intended and on any appropriate precautions to be observed.</td></tr>
<tr><td></td><td>51.2</td><td>Adequacy of information (such as ingredients and nutrition information) and its intelligibility to the intended user are of particular importance. This includes the need to recognise possible literacy and language problems, stock rotation and effective product life, and appropriateness of presentation and packaging (e.g. in-pack microwaveability or product stability during prolonged maintenance at serving temperature and relative humidity).</td></tr>
<tr><td></td><td>51.3</td><td>Where a catering/food service organisation is preparing food by factory-style cook-freeze or cook-chill processing in the United Kingdom (UK), reference should also be made to the Food Standards Agency (FSA) guidelines or further information such as Campden BRI Guidelines.</td></tr>
<tr><td></td><td>51.4</td><td>The interaction effects between product, environment and equipment are especially pertinent in operations involving vending machines. In the manufacture of products for these purposes, the manufacturer should ensure awareness of such potential hazards as within-machine environment, hygiene and cleaning needs, product flow properties, variability of throw or dispensation, as well as product and machine interactions and interactions with other products or ingredients.</td></tr>
</table>

Food & Drink – Good Manufacturing Practice: A Guide to its Responsible Management, Seventh Edition.
The Institute of Food Science & Technology Trust Fund.
© 2018 John Wiley & Sons Ltd. Published 2018 by John Wiley & Sons Ltd.

51.5　The requirements for vending operations may call for particular product performance standards, for example dispersion at sub-scalding temperatures or interchangeability with competitors' products in an identical vending situation.

51.6　Reference may also be made to the Automatic Vending Association (AVA), who have developed their own AVA Quality System that their members must comply with. Further details can be found at http://the-ava.com and the European Vending Association (EVA) at http://www.vending-europe.eu/eva/home.html.

Complaints　51.7　Manufacturers should develop appropriate systems to handle, record and respond to complaints from caterers/food service organisations, and from their own customers, as these will not normally fit into the system for dealing with direct complaints from consumers who have purchased retail products (see Chapter 27).

52 THE USE OF FOOD ADDITIVES AND PROCESSING AIDS

Principle

The manufacturer should satisfy him/herself that the use of a processing aid serves an essential functional need in manufacture, and that the inclusion of any additive serves an essential technical need or contributes significantly to customer quality requirements, or both, and does not present a food safety risk. The use of any food additive or a processing aid must comply with the laws and specific regulations of the country for which the food is intended.

Food Additives

52.1 Governments and their expert advisers must consider both the toxicological evaluation and the need for the substance in question, and establish safety-in-use criteria before approving an additive and review such approval on a continuing basis. These principles already apply in the United Kingdom (UK) and according to their legislation in many other countries. Food additives are substances that are intentionally added to foods in order to perform a certain function, such as to sweeten, preserve, colour, influence consistency, stabilise and extend shelf life. European Union (EU) legislation defines a food additive as 'any substance not normally consumed as a food in itself and not normally used as a characteristic ingredient of food, whether or not it has nutritive value, the intentional addition of which to food for a technological purpose in the manufacture, processing, preparation, treatment, packaging, transport or storage of such food results, or may be reasonably expected to result, in it or its by-products becoming directly or indirectly a component of such foods'.

52.2 The use of an additive must not mislead consumers as to the quality of a product. This necessitates appropriate labelling that includes information about additives present given in accordance with the regulations of the country for which the food is intended. It also implies that the labelling and advertising language or pictorial representation should not be used to create a false or misleading impression.

52.3 Additives should not be used to permit or to disguise the effect of faulty processing.

52.4 **Regulation EC No. 1333/2008[1] of the European Parliament and of the Council of 16 December 2008 on food additives** sets out the rules on food additives, providing definitions, an outline of conditions of use, and labelling guidelines. Annex I addresses the technological functions of food additives, Annex II provides a list of

[1] http://eur-lex.europa.eu/legal-content/EN/TXT/?uri=CELEX:32008R1333.

Food & Drink – Good Manufacturing Practice: A Guide to its Responsible Management, Seventh Edition.
The Institute of Food Science & Technology Trust Fund.
© 2018 John Wiley & Sons Ltd. Published 2018 by John Wiley & Sons Ltd.

food additives approved for use in food additives and conditions of use, Annex III lists food additives approved for use in food additives, food enzymes and food flavourings, and their conditions of use, Annex IV addresses traditional foods for which certain Member States may continue to prohibit the use of certain categories of food additives and Annex V gives additives labelling information for certain food colours. The list of authorised food additives approved for use in food additives, enzymes and flavourings can be found in the Annex of **Commission Regulation (EU) No. 1130/2011**[2] **of 11 November 2011 amending Annex III to Regulation (EC) No. 1333/2008 of the European Parliament and of the Council on food additives** by establishing an EU list of substances approved for use such as food additives, food enzymes, food flavourings and nutrients. Food additives must comply with specifications that include information that adequately identifies the food additive, including origin, and the acceptable criteria of purity. **Commission Regulation (EU) No. 231/2012**[3] of 9 March 2012 lays down specifications for food additives listed in Annexes II and III to **Regulation (EC) No. 1333/2008 of the European Parliament and of the Council Text**. Further guidance documents are provided by the EU online and can be downloaded.[4] An additive database is also provided on the EU food additives website that serves as a tool to inform manufacturers about the food additives approved for use and their conditions of use.[5]

52.5   In the UK and other EU Member States, where a maximum level of usage of any substance is specified in the appropriate directives and regulations, the level must not be exceeded, but, additionally, the level of usage of a specific substance should not exceed the lowest level that produces the required effect when used efficiently in good manufacturing practice (GMP). It is important to note that some additives, such as sulphur dioxide, are also classed as allergens, which require specific labelling if they are present in a food product (see Chapter 8). In the UK, the Food Standards Agency (FSA) Food and Feed Law Guide[6] identifies the current EU and national legislation with regard to food additives. The FSA has also produced a document called *Food Additives Legislation Guidance to Compliance* (October 2015),[7] which covers both food additives and processing aids. The FSA has a designated site which provides details on additives (see https://www.food.gov.uk/science/additives). Processing aids, including filtration aids and release agents are excluded from the scope of **Regulation No. 1333/2008.**

[2] http://eur-lex.europa.eu/legal-content/EN/TXT/?uri=CELEX:32011R1130.
[3] http://eur-lex.europa.eu/legal-content/EN/TXT/?uri=CELEX:32012R0231.
[4] https://ec.europa.eu/food/safety/food_improvement_agents/additives/eu_rules_en.
[5] https://webgate.ec.europa.eu/foods_system/main/?sector=FAD&auth=SANCAS.
[6] https://www.food.gov.uk/sites/default/files/food_feed_law_guide_dec2016.pdf.
[7] https://www.food.gov.uk/sites/default/files/multimedia/pdfs/guidance/food-additives-legislation-guidance-to-compliance.pdf.

Processing Aids   52.6   In the UK there is no national legislation on processing aids nor a legally defined list of approved processing aids. The term 'processing aid' means 'any substance which: is not consumed as a food by itself; is intentionally used in the processing of raw materials, foods or their ingredients, to fulfil a certain technological purpose during treatment or processing; and may result in the unintentional but technically unavoidable presence in the final product of residues of the substance or its derivatives provided they do not present any health risks and do not have any technological effect on the final product.'[7] It is important to consider processing aids that are derived from materials that can cause allergic reactions in sensitive individuals, for example isinglass derived from fish and used as a clarification aid in brewing and wine production (see Chapter 8).

53   RESPONSIBILITIES OF IMPORTERS

Principle

There is a responsibility on the importer of foods and drinks to satisfy her/himself and the appropriate authorities that the imported products have been produced in accordance with the principles of good manufacturing practice (GMP) and that they, and their mode of packaging and distribution, comply with the relevant legislative requirements at the points of import and sale.

53.1   In all respects, the importer should ensure that the requirements outlined in this Guide are met as though she/he were, her/himself, the producer of the products (for the import of food into the UK, attention is drawn to the due diligence aspects of the **Food Safety Act 1990** and subsequent legislation).

53.2   The importer should ensure that all imports are obtained against a clear and legally valid product specification. Attention is drawn to a United Kingdom (UK) legal requirement to obtain appropriate certification in certain instances. Wherever possible, inspection should be carried out at the point of origin to ensure that the agreed specification is met.

53.3   The goods on receipt in the country of destination should be subjected to a food safety and quality control evaluation that should take due account of any changes or damage that might have occurred in transit, and should reflect the confidence level established in the competence of the supplier and the current and historic availability of data on product safety, quality and integrity from that origin. On importation, all the relevant legislative requirements, including those pertaining to labelling and metrology, must be met.

53.4   Consideration should be given to the requirements of local health authorities and port health authorities, and the possibility of problems arising from their needs to examine and clear shipments. Close liaison with officers is to be commended, particularly with regard to sampling, clearance of perishable goods, identification requirements and any special needs relating to the product in question.

53.5   The importer should ensure adequate liaison with the manufacturer on matters relating to legislative or other changes in requirements for the products, and exchange information resulting from complaints, the needs of his/her own customers and the ultimate consumers.

Food & Drink – Good Manufacturing Practice: A Guide to its Responsible Management, Seventh Edition.
The Institute of Food Science & Technology Trust Fund.
© 2018 John Wiley & Sons Ltd. Published 2018 by John Wiley & Sons Ltd.

54   EXPORT

Principle

The manufacture of foods for export purposes requires general compliance with the principles embodied throughout this Guide, namely those of meeting specific legislation and/or regulatory obligations, and of ensuring that packaging and shipping arrangements are adequate and suitable. It is the responsibility of the producer, unless manufacturing for a specific contract on behalf of the importer or ultimate customer, to be aware of and follow the pertinent rules relating to export from the country of origin and to acceptance in the country of destination.

54.1   Licensing or inspection of the producing plant may be required so that formal approval can be given prior to seeking to export product, as may certification of materials or processes. In some cases, inspection or enforcement may lie with the authorities from the country of destination, or their agents, while in others, national or local authorities may carry out an acceptance function.

54.2   Due regard should be paid to the requirements of port health authorities and customs both in the United Kingdom (UK) and the importing country.

54.3   Attention should be paid to the packaging and shipping requirements, particularly with regard to dimensions, weights, stowage, protection, hygiene and contamination risks. Where appropriate, clear instructions should be provided on the packaging and/or in accompanying documentation on any special treatment or precautions required and procedures to adopt in the event of breakage, accident or delay.

54.4   Due attention to use of appropriate languages and international symbols should be given in relation to transit arrangements and ultimate destination.

54.5   In the manufacture of products for export, due regard should be paid to the suitability of the product formulation and packaging for the environmental circumstances likely to be encountered *en route* and at the ultimate destination.

Food & Drink – Good Manufacturing Practice: A Guide to its Responsible Management, Seventh Edition.
The Institute of Food Science & Technology Trust Fund.
© 2018 John Wiley & Sons Ltd. Published 2018 by John Wiley & Sons Ltd.

PART III – MECHANISMS FOR REVIEW OF THIS GUIDE

This Guide is updated from time to time to incorporate any changes found to be necessary in the light of experience, changing legislation or advances in food science and technology and to correct errors or omissions and so on.

Food & Drink – Good Manufacturing Practice: A Guide to its Responsible Management, Seventh Edition.
The Institute of Food Science & Technology Trust Fund.
© 2018 John Wiley & Sons Ltd. Published 2018 by John Wiley & Sons Ltd.

APPENDIX I

DEFINITION OF SOME TERMS USED IN THIS GUIDE

Adulteration	Intentionally adding extraneous, improper or inferior ingredients to a food product.
Analytical Method	A detailed description of the procedures to be followed in performing tests for assessing conformity with the specification.
Aseptic Processing	A production process where a commercially sterile product (see Commercial Sterility), otherwise known as aseptic product, is packed into a container that has been independently and previously sterilised in a way that prevents contamination and maintains full sterility. Examples include Tetra Pak technology and Bag-in-Box technology.
Authentic	Products that demonstrate a given connection to a recipe, location or social characteristic.
Authenticity	The innate quality of being authentic, genuine and of undisputed origin.
Batch	The quantity of material that has been produced during a defined period of manufacture. A 'batch' may actually have been produced by a batch-wise process, or may correspond to a particular time duration during the run of a continuous process.
	Or an identifiable amount of product or material that has undergone a given treatment or process at a given time in a given location.
Batch Manufacturing Record	A document stating the materials used and operations carried out during the manufacture of a given batch, including details of in-process controls and the results of any corrective action taken. It should be based on the master manufacturing instructions, and be compiled as the manufacturing operation proceeds.
Batch Number	A unique combination of numbers or letters, or both, used to identify a batch and permit its history to be traced.
Botulinum Cook	The heat treatment given to a low-acid canned food (having a pH higher than 4.5) sufficient to inactivate 10^{12} spores of *Clostridium botulinum*. This heat treatment is called the F_o value, and it is equivalent to a process of 3 minutes at 121°C.
Bulk Product	Any product that has completed all processing stages up to, but not including, packaging (not applicable to those products where processing takes place inside the container and the latter is itself therefore part of the processing).

Food & Drink – Good Manufacturing Practice: A Guide to its Responsible Management, Seventh Edition.
The Institute of Food Science & Technology Trust Fund.
© 2018 John Wiley & Sons Ltd. Published 2018 by John Wiley & Sons Ltd.

Chill Chain	An organised system governing the conditions under which chilled foods are stored and handled by the producer, distributor and retailer. The conditions, as set out in Chapter 44, are those which ensure that temperatures maintained during storage, distribution and sale are those consistent with maintenance of quality and safety.
Chilled Foods	Perishable foods that, to extend the time during which they remain wholesome, are kept within controlled and specified ranges of temperature above their freezing points and normally below 8°C.
Cold Chain	An organised system governing the conditions under which frozen foods are stored and handled by the producer, distributor and retailer. The conditions, as set out in Chapter 45, are those which ensure that temperatures maintained during storage, distribution and sale are those consistent with maintenance of quality and safety.
Commercial Sterility	A term common in the canning industry meaning the condition achieved by the application of heat that renders the processed product free from viable micro-organisms, including those of known public health significance, capable of growing in the food at the temperatures at which the food is likely to be held during distribution and storage.
Contract Manufacture	Manufacture or partial manufacture ordered by one person or organisation (the contract giver) and carried out by a separate person or organisation (the contract acceptor).
Corrective action	The action taken on the identification of non-conformance in terms of ingredients, products, behaviours or activities with specified documented requirements.
Countermeasure	The action taken by an individual, organisation or other body to counteract or offset a given danger or threat.
Critical Control Point	CCP; a material, or a location, or a practice, or a procedure, or a process stage where loss of control would result in an unacceptable food safety risk.
Cybersecurity	The measures taken to protect a computer system or individual appliance against an intentional malicious target attack and/or unauthorised access and unintentional or accidental access.
Dehydrated Food	Food or food products from which all but a small percentage of the water has been removed under controlled conditions.
Detergent	A chemical or mixture of chemicals that is used to remove soiling or grease from a surface, leaving it accessible to the action of disinfectants.
Disinfection	Defined by British Standard 5283:1986 as 'the destruction of microorganisms, but not usually bacterial spores; it may not kill all microorganisms but reduces them to a level which is neither harmful to health, nor the quality of perishable foods'.

Documentation	All the written production procedures, instructions and records, quality control procedures and recorded test results involved in the manufacture of a product.
Economically Motivated Adulteration	The process of adulteration with the specific intent of reducing the cost of production or misrepresenting a food to increase profits.
Extrinsic	A property derived from outside of the food product or ingredient; for example an extrinsic hazard could arise from contamination from the environment, equipment, people or pests or extrinsic quality attribute could describe the way the product or ingredients have been grown or produced, for example, free range, organic (see Intrinsic).
Finished Product	A product that has undergone all stages of manufacture and packaging.
Food Allergen	A food substance that, in some sensitive individuals, causes an immune response causing bodily reactions resulting in the release of histamine and other substances into the tissues from the body's mast cells in the eyes, skin, respiratory system and intestinal system. Allergic reactions may range from relatively short-lived discomfort to anaphylactic shock and death.
Food Control	The Institute of Food Science & Technology uses the term 'food control' to describe a comprehensive quality and safety system involving hazard analysis critical control point or HACCP-linked food safety, integrity and quality assurance and food safety, integrity and quality control.
Food Crime	Dishonesty relating to the production or supply of food that is either complex or likely to be seriously detrimental to consumers, businesses or the overall public interest (see 5.1).
Defence	The activities required to protect a food product or indeed a food supply chain from intentional or deliberate acts of contamination and includes ideological or malicious threat (see 5.4).
Food Fraud	A dishonest act or omission, relating to the production or supply of food, which is intended for personal gain or to cause loss to another party (see 5.1).
Food Hygiene	All environmental factors, practices, processes and precautions involved in protecting food from contamination by any agency, and preventing any organism present from multiplying to an extent that would expose consumers to risk or result in premature spoilage or decomposition of food.
Food Integrity	Concerned with the nature, substance and quality and safety of food, as well as other aspects of food production such as the way food has been 'sourced, procured, and distributed and being honest about those areas to consumers' (see 5.1).

Food Integrity Management System	FIMS; a documented set of policies, procedures and associated documentation that combines to form a comprehensive system that effectively manages food integrity in terms of product, process, data and people.
Food Poisoning	Illness associated with consumption of food that has been contaminated, particularly with harmful microorganisms or their toxins.
Food Safety Management System	FSMS; a documented set of policies, procedures and associated documentation that combines to form a comprehensive system that effectively manages food safety.
Food Spoilage	The deterioration of food, including that caused by the growth of undesirable micro-organisms to high levels, which may result in fermentation, mould growth and development of undesirable odours and flavours.
Frozen Foods	Foods preserved by freezing and storing at temperatures low enough to inhibit the growth of micro-organisms and to retard chemical and physical reactions to a negligible rate.
Functional	Fulfilling a specific physical, chemical or biological function.
Functional Foods	The term is one of the marketing-coined names (others are 'nutraceuticals' and 'designer foods') to categorise foods that are considered or claimed to offer specific health benefits over and above those provided by recognised nutrients, while avoiding the requirement to be licensed medicines.
Genetic Modification	GM; the process of making changes to the genes of an organism (whether an animal or plant organism or a microorganism). Genetic changes occur spontaneously in nature over a long period of time, but they may be produced intentionally either by traditional methods of selective breeding of animals and plants or by modern methods of removal or insertion of genes.
Genetically Modified Organism	GMO; descriptive of an organism undergoing genetic modification or of an organism resulting from genetic modification (see above).
Good Manufacturing Practice	GMP; that combination of manufacturing and quality control procedures aimed at ensuring that products are consistently manufactured to their designated specifications.
HACCP Plan	A document that is prepared using the seven principles of hazard analysis critical control point, as defined in Codex Alimentarius, in order to identify realistic food safety hazards, the points at which they could arise in the manufacturing process (including materials intake and transport to the consumer) and the means for their effective control.
Hazard	A property of a system, operation, material or situation that could, if uncontrolled, lead to an adverse consequence.

340

Hazard Analysis	Preparation of a list of the steps in a manufacturing process (best done by preparing and verifying a process flow chart), identification of points at which hazards could arise and then an assessment of the nature and potential seriousness of each hazard so as to establish critical control point(s) and the means for their effective control.
Hazard Analysis and Operability Study	HAZOP; a systematic structured approach to questioning the sequential stages of a proposed operation in order to optimise the efficiency and the management of risk. Thus, the application of HAZOP to the design of a proposed food-related operation should result in a system in which as many critical control points as possible have been eliminated, making hazard analysis critical control point during subsequent operations much easier to carry out.
Hazard Analysis Critical Control Point	HACCP; a systematic preventive food safety tool designed to assist manufacturing organisations to develop an appropriate and effective food safety management system.
High-care Area	HCA; an area designed to a high standard of facility specification and hygienic design where practices relating to personnel, ingredients, equipment and environment are managed to minimise microbial contamination of a ready-to-eat or ready-to-reheat product containing uncooked ingredients
High-risk Area	HRA; an area designed to a high standard of facility specification and hygiene design where practices relating to personnel, ingredients, equipment and environment are managed to minimise microbial contamination of a ready-to-eat or ready-to-reheat product comprising only cooked ingredients.
Identity Preserved Material	A material or food product that is traceable to a known source or specific method of growing or food production (see Provenance), for example kosher, halal, organic or farm assured.
Ingredients	All materials, including starting materials, processing aids, additives and compounded foods, that are included in the formulation of the product.
In-process Control	A system of checks made and actions taken during the course of manufacture to ensure that materials at any stage comply with the specification for that stage, and that the processing and processing environment comply with the conditions stated in the master manufacturing instructions.
Intermediate Material	A partly processed material that must undergo further processing before it becomes a bulk product or a finished product.
Intrinsic	An inherent component of a food. Intrinsic food safety hazards are derived from the product, for example bones in fish and stones in fruit. Control measures are therefore product specific and can include declarations on the packaging of the finished product. Intrinsic quality relates to characteristics of the food itself for example shape, colour or physio-chemical characteristics.

Irradiated	Having been subjected to ionising radiation.
Lot	ISO 20005:2007 defines a lot as 'a set of units of a product which has been produced and/or processed or packaged under similar circumstances'. Alternatively, vocabulary is used in the manufacturing sector such as *lot* (group of items defined by the manufacturer; see also 14.9), *trade unit* (quantity defined at its lowest repeatable unique unit, e.g. jar, box, pack, bag) and *logistic unit* (quantity defined at its lowest repeatable unique unit for storage or transportation, e.g. tray, box, pallet or container).
Low-risk Area	LRA; an area where good manufacturing practice standards are in place as described within this publication but the area and the practices have not been specifically designed to minimise microbial contamination, for example raw material intake, storage areas of ready-to-cook foods and packaged product where the product is fully enclosed.
Manufacture	The complete cycle of production of a food or drink product from the acquisition of all materials through all stages of subsequent processing, packaging and storage to the despatch of the finished product.
Master Manufacturing Instructions	A document or documents identifying the raw materials, with their quantities, to be used in the manufacture of a product, together with a description of the manufacturing operations and procedures, including identification of the plant and facilities to be used, processing conditions, in-process controls, packaging materials to be used and instructions for the removal of the finished product to storage.
Monitoring	The process of undertaking a prescheduled sequence of observations or measurements to assess whether a product, ingredient, process, procedure, prerequisite programme or critical control point is adequately controlled in order to consistently manufacture safe and legal food of the required quality.
Non-conformance/ Non-compliance	A failure to comply with an element of the food safety management system, food integrity management system, quality management system or hazard analysis critical control point plan. The failure can be classed as major or minor. The term 'critical' is sometimes used to highlight a major non-conformance that has been identified that could lead to a food safety or legality issue.
Novel (Food, Process)	A food or food ingredient produced from raw material that has not hitherto been used (or has been used only to a small extent) for human consumption in the area of the world in question, or that is produced by a new or extensively modified process not previously used in the production of food.
Nutraceutical	See Functional Foods.

Packaging	Any container or material used in the packaging of a product. This may include materials in direct contact with the product, printed packs, including labels, carrying statutory and other information, and other packaging materials, including outer cartons or delivery cases. These categories are, of course, not necessarily mutually exclusive.
Prerequisite Programme	PRP; an element of the food safety management system (FSMS) food integrity management system and/or quality management system that has been adopted by the manufacturer in order to effectively manage food safety, legality and compliance with quality specifications. The existence and effectiveness of prerequisite programmes should be considered during the development of the FSMS and hazard analysis critical control point plan and in the determining of critical control points.
Preservation Index	A term deriving from the pickles and sauces industry to designate the percentage of acetic acid contained in the total volatile constituents of a product or ingredient, thus indicating probable microbial stability.
Preventive Action	An action that is adopted in order to address a weakness in a food safety management system, food integrity management system and/or quality management system that has not to date been responsible for non-conformance, but if not addressed could be in the future.
Processed	Having been subjected to treatment designed to change one or more of the properties (physical, chemical, microbiological, sensory) of food.
Processing Aid	The UK Food Labelling Regulations 1996 define processing aid as 'any substance not consumed as a food by itself, intentionally used in the processing of raw materials, foods or their ingredients, to fulfil a certain technological purpose during treatment or processing, and which may result in the unintentional but technically unavoidable presence of residues of the substance or its derivatives in the final product, provided that these residues do not present any health risk and do not have any technological effect on the finished product'. It follows that a processing aid is an additive that facilitates processing without significantly influencing the character or properties of the finished product.
Provenance	Relates to not only the geographic elements of where the ingredients and the final food are grown, processed and finally manufactured, but also how that food is produced and whether the methods of production and processes employed comply with certain standards and protocols.
Quality Assurance	See Chapter 2.
Quality Control	See Chapter 2.
Quality Management System	QMS; the organisational framework of policies, procedures, associated documentation and resources needed to implement the strategy required to consistently deliver product that is within predetermined specifications.

Quarantine	The status of any materials or product set aside while awaiting a decision on its suitability for its intended use or sale.
Radiation Dose	Doses of radiation are defined in terms of the energy absorbed by the substance irradiated. The unit for radiation dose is the gray (Gy), which is defined as the dose corresponding to the absorption of 1 joule per kilogram of the matter through which the radiation passes [1 kilogray (kGy) = 1000 Gy].
Raw Material	Any material, ingredient, starting material, semi-prepared or interme- diate material, packaging material and so on used by the manufacturer for the production of a product.
Ready to Cook	RTC; food designed to be given a heat process by the consumer that will deliver a 6-log kill with respect to vegetative pathogens (a minimum process equivalent to 70°C for 2 minutes) throughout all components.
Ready to Eat	RTE; food intended by the producer or the manufacturer for direct human consumption without the need for cooking or other processing effective to reduce to an acceptable level or eliminate microorganisms of concern (i.e. cold eating).
Ready to Reheat	RTRH; food manufactured in a high-care area or high-risk area that has been designed to be reheated by the final consumer.
Revalidation	The process by which an element of the food safety management system, food integrity management system, quality management system or hazard analysis critical control point plan is reassessed to ensure its continuing ability to deliver food safety, integrity, quality or legal compliance objectives.
Reworking	The process of taking food that does not meet specification and reprocessing, resorting or otherwise handling it in order to address the non-conformance so that it will then meet specification.
Risk	The probability that a particular adverse consequence results from a hazard within a stated time under stated conditions.
Risk Assessment	The management process of identification, evaluation and estimation of the levels of risk associated with a food safety hazard, situation or process or procedural failure. Further categorisation of the risk identified is by determining the likelihood and the severity of occurrence and determining the acceptable level of risk to the consumer. This can include the acceptable level of a component in a food product, for example a pesticide residue, or absence versus presence of a foreign body.
Root Cause Analysis	The structured management approach that identifies the factors that resulted in non-conformance in order to determine the most appropriate corrective or preventive action. The factors that could be considered include the actual nature of the non-conformance, the magnitude (major or minor) and the consequences of the problem in order to identify the actions, conditions or behaviours that need to be changed to prevent reoccurrence and/or other similar problems from occurring.

Specification	A document giving a description of material, machinery, equipment, process or product in terms of its required properties or performance. Where quantitative requirements are stated, they are in terms of either limits or standards with permitted tolerances.
Starting Material	See Raw Material.
Threat	A threat is defined by PAS 96:2017 as 'something that can cause loss or harm which arises from the ill-intent of people'.
Threat Analysis Critical Control Point	TACCP; PAS 96:2017 describes threat analysis critical control point as a 'systematic management of risk through the evaluation of threats, identification of vulnerabilities, and implementation of controls to materials and products, purchasing, processes, premises, distribution networks and business systems by a knowledgeable and trusted team with the authority to implement changes to procedures'.
Traceability	Regulation EC/178/2002 defines traceability as the ability to trace and follow a food, feed, food-producing animal or substance intended to be, or expected to be, incorporated into a food or feed through all stages of production, processing and distribution.
Traceable Resource Unit	TRU; a unique batch of material or product that, using a specific set of traceability characteristics, is readily distinguishable from other batches of material.
Validation	The process of obtaining of evidence that the elements of the food safety management system, food integrity management system, quality management system or hazard analysis critical control point plan are effective at delivering food safety, integrity, quality or legal compliance objectives.
Verification	The process of developing procedures, assessments and other evaluations, including auditing, that is in addition to monitoring to determine compliance with the food safety management system, the food integrity management system, the quality management system and hazard analysis critical control point plans.

APPENDIX II

ABBREVIATIONS USED IN THIS GUIDE

Note: In some instances, an abbreviation of the former name of an organisation is included where a document referred to was issued by, and bears the former name of, that organisation.

AAS	Atomic absorption spectrometry
ACNFP	Advisory Committee on Novel Foods and Processes (UK)
AIDC	Automatic identification data capture
ALOP	Acceptable level of protection
APC	Aerobic plate count
ATP	Adenosine triphosphate
AVA	Automatic Vending Association [formally the Automated Vending Association of Britain (AVAB)]
B2B	Business to business
B2C	Business to consumer
BCMS	Business continuity management plan
BPCA	British Pest Control Association
BRC	British Retail Consortium
BRDO	Better Regulation Delivery Office
BS	British Standard
BSI	British Standards Institution
CAC/RCP	Codex Alimentarius Commission/Recommended Code of Practice
CAC/RS	Codex Alimentarius Commission/Recommended Standard
CAPA	Corrective and preventive action

Food & Drink – Good Manufacturing Practice: A Guide to its Responsible Management, Seventh Edition.
The Institute of Food Science & Technology Trust Fund.
© 2018 John Wiley & Sons Ltd. Published 2018 by John Wiley & Sons Ltd.

CARVER+ SHOCK	Criticality, accessibility, recuperability, vulnerability, effect, recognisability, plus Shock
CCP	Critical control point
CCTV	Closed-circuit television
CEN	Comité Européene de Normalisation
CFA	Chilled Food Association
CIP	Cleaning in place
COO	Country of origin
COSHH	Control of Substances Hazardous to Health
CPD	Continuous professional development
Cp–PyMS	Curie point pyrolysis coupled to mass spectrometry
CQP	Critical quality point
CSF	Critical success factor
CSP	Corporate social performance
CSR	Corporate social responsibility
DEFRA	Department of Environment, Food and Rural Affairs (UK)
DNA	Deoxyribonucleic acid
EAS	Electronic article surveillance
EC	European Community
EDI	Electronic date interchange
EDP	Electronic data processing
EFK	Electric fly killer (or insectocutor)
EFSA	European Food Safety Authority
EHO	Environmental health officer
EID	Electronic identification
ELISA	Enzyme linked immuno-sorbent assays

EMA	Economically motivated adulteration
EN	Denotes a regional standard intended to be used in the European Union
ETI	Ethical trading initiative
EU	European Union [formerly European Community (EC) and originally Economic Community (EEC)]
EUPFN	European Union Protected Food Name Scheme
EVA	European Vending Association
FAIR	Food Adulteration Incidents Registry
FALCPA	Food Allergen Labelling and Consumer Protection Act 2004
FAO/WHO	Food and Agriculture Organisation/World Health Organisation (United Nations)
FAPAS	Food Analysis Performance Assessment Scheme
FCRA	Food crime risk assessment
FDA	Food and Drug Administration (USA)
FDO	Food distribution organisation
FIC	Food information (to) consumers
FIFO	First-in, first-out
FIMS	Food integrity management system
FLCP	Food Law Code of Practice
FLT	Fork lift truck
FMA	Food Machinery Association (UK)
FMEA	Failure mode and effects analysis
FMF	Food Manufacturers' Federation (UK)
FRC	Free residual chlorine
FSA	Food Standards Agency
FSMS	Food safety management system

FSO	Food safety objective
FT-MIRS	Fourier transform mid-infrared spectroscopy
GAP	Good agricultural practice
GC	Gas chromatography
GDP	Good distribution practice
GFSI	Global food safety initiative
GHP	Good hygienic practice
GIS	Geographic information systems
GLAA	Gangmaster and Labour Abuse Authority (formerly Gangmaster Licensing Authority)
GLP	Good laboratory practice
GM	Genetically modified
GMO	Genetically modified organism
GMP	Good manufacturing practice
GPS	Global positioning system
GRI	Global reporting initiative
HACCP	Hazard analysis critical control point
HAZOP	Hazard analysis and operability study
HCA	High-care area
HMSO	Her Majesty's Stationery Office (now The Stationery Office)
HPA	Health Protection Agency
HPLC	High-performance liquid chromatography
HR	High risk
HRA	High-risk area
HR-MS	High-resolution mass spectroscopy
HSE	Health and Safety Executive

ICCT	International cold chain technology
ICMSF	International Committee for Microbiological Specifications for Foods
IEC	International Electrotechnical Commission
IFST	Institute of Food Science & Technology (UK)
IOSH	Institute of Occupational Safety and Health (IOSH)
IPPC	Integrated pollution prevention and control
IRMS	Isotope ratio mass spectrometry
ISO	International Organization for Standardization
ISO/TS	International Organization for Standardization/Technical Specification
JECFA	Joint FAO/WHO Expert Committee on Food Additives
KPI	Key performance indicator
LA	Local authority
LACORS	Local Authorities Coordinators of Regulatory Services (UK) [formerly LACOTS (Local Authorities Coordinating Body on Trading Standards)]
LOAEL	Lowest observable adverse effect level
LR	Low risk
LRA	Low-risk area
MAFF	Ministry of Agriculture, Fisheries and Food (UK)
MANCP	Multi-Annual National Control Plan
MAP	Modified atmosphere packaging
MIR	Mid infrared
MIRS–NIRS	Mid and near infrared spectroscopy
MITE	Measuring inspection and testing equipment
MRA	Microbiological risk assessment

MS	Mass spectroscopy
NAMAS	National Measurement Accreditation Service (see UKAS)
NFCU	National Food Crime Unit
NIR	Near infrared
NMR	Nuclear magnetic resonance
NMR/ MSIR	Nuclear magnetic resonance coupled with mass spectrometry of isotopic ratio
NOAEL	No observable adverse effect level
NPTA	National Pest Technicians Association
OCR	Optical character recognition
OECD	Organisation for European Co-operation and Development
PAS	Publicly available specification
PCR	Polymerase chain reaction
PCR-LP	PCR-length polymorphism
PDCA	Plan, do, check, act
PDO	Protected designation of origin
PGI	Protected geographic indication
PHE	Public Health England
PPE	Personal protective equipment
PR	Pathogenesis-related
PRP	Prerequisite programme
QFF	Quick frozen foodstuffs
RASFF	Risk Assessment Food and Feed (EU food and feed alert forum)
RCP	Recommended Code of Practice
RFID	Radio frequency identification

RIDDOR	Reporting of Injuries, Disease and Dangerous Occurrences Regulations 2013
RS	Remote sensing
RTC	Ready to cook
RTE	Ready to eat
RTP	Returnable transit packaging
RTRH	Ready to reheat
Q	Qualitative
QMS	Quality management system
QR	Quick response (code)
QRA	Quantitative risk assessment
QUID	Quantitative ingredient declaration
SA	Social accountability
SEDEX	Supplier Ethical Data Exchange
SFBB	Safer Food Better Business
SI	Statutory Instrument (UK)
SMART	Specific, measureable, achievable, realistic, time-based
SMETA	Sedex Members Ethical Trade Audit
SMS	Security management system
SNIF-NMR	Site-specific natural isotope fractionation by nuclear magnetic resonance
SNP	Single nucleotide polymorphism
SPC	Statistical process control
SPS	Sanitary and phytosanitary
SQ	Semi-quantitative
STR	Short tandem repeats
TACCP	Threat analysis critical control point

TRU	Traceable resource unit
TSG	Traditional speciality guaranteed
TSO	Trading standards officer
UK	United Kingdom
UKAS	UK Accreditation Service (formerly NAMAS, q.v.)
USA	United States of America
USDA	United States Department of Agriculture
UV	Ultraviolet
VACCP	Vulnerability assessment critical control point
VAS	Vulnerability assessment software
VITAL	Voluntary incidental trace allergen labelling
VR	Verification risk
WHO	World Health Organization
WTO	World Trade Organization
XML	Extensible markup language

APPENDIX III

LEGISLATION AND GUIDANCE

There is a vast number of guides and industry codes of practice issued by a variety of bodies, some of which are referenced in the chapters. Previous editions of this Guide have included a comprehensive listing of such guides and codes. These will not be repeated in this appendix; instead, a list of websites is identified where further information can be accessed.

The *Food Law Guide* is issued by the Food Standards Agency (FSA) and is an invaluable listing of UK Food Laws and Regulations, with brief descriptive outlines and indication of European Union (EU) Directives to which each gives effect in the UK. The guide is updated quarterly and is available on the FSA website. To take account of legislative provisions, reference should be made to the actual texts of the relevant laws and regulations, purchasable from The Stationery Office or any of its book-shops (http://www.tso.co.uk).

Websites

Automatic Vending Association
http://the-ava.com

British Frozen Food Federation
http://www.bfff.co.uk

British Pest Control Association
http://www.bpca.org.uk

British Retail Consortium
https://brc.org.uk

British Standards Institution
http://www.bsigroup.co.uk

Campden BRI
http://www.campden.co.uk

Chilled Food Association
http://www.chilledfood.org

Codex Alimentarius Standards
http://www.fao.org/fao-who-codexalimentarius/en/

Food & Drink – Good Manufacturing Practice: A Guide to its Responsible Management, Seventh Edition.
The Institute of Food Science & Technology Trust Fund.
© 2018 John Wiley & Sons Ltd. Published 2018 by John Wiley & Sons Ltd.

Department of Food, Environment & Rural Affairs (UK)
https://www.gov.uk/government/organisations/department-for-environment-food-rural-affairs

Department of Health & Social Care
https://www.gov.uk/government/organisations/department-of-health-and-social-care

European Union: European Food Safety Authority
http://www.efsa.europa.eu/

European Vending Association
http://www.eva.be

Food and Agriculture Organization of the United Nations
http://www.fao.org

Food and Drink Federation
https://www.fdf.org.uk

Food Standards Agency
http://www.food.gov.uk

Food Standards Agency National Food Crime Unit
https://www.food.gov.uk/enforcement/the-national-food-crime-unit

Global Food Safety Initiative
https://www.mygfsi.com

Institute of Food Science & Technology Information Statements
https://www.ifst.org/knowledge-centre/information-statements

Institute of Food Science & Technology
http://www.ifst.org

International Commission on Microbiological Specifications for Foods
http://www.icmsf.org/index.html

National Pest Technicians Association
http://www.npta.org.uk

Public Health England
https://www.gov.uk/government/organisations/public-health-england

The Royal Society
https://royalsociety.org

The Stationery Office
http://www.tso.co.uk

World Health Organization
http://www.who.int/

ADDITIONAL REFERENCES

This appendix provides some references to standards and topics relevant to good manufacturing practice (GMP) and its management. The sheer volume of the publications available makes it impractical for the list to be exhaustive either in scope or within subject.

BRC Global Standard for Food Safety: Issue 7 The British Retail Consortium, The Stationery Office, July 2015. Other guidance available at the TSO website.

Standards referred to in the Guide include:

BS EN 1276:2009. Chemical disinfectants and antiseptics. Quantitative suspension test for the evaluation of bactericidal activity of chemical disinfectants and antiseptics used in food, industrial, domestic, and institutional areas. Test method and requirements (phase 2, step 1). May 2010.

BS EN 13697:2015. Chemical disinfectants and antiseptics. Quantitative non-porous surface test for the evaluation of bactericidal and/or fungicidal activity of chemical disinfectants used in food, industrial, domestic and institutional areas. Test methods and requirements without mechanical action. April 2015.

BS EN 1499-2013. Chemical disinfectants and antiseptics – Hygienic hand-wash – Test method and requirements (phase 2/step 2). May 2013.

BS EN 1500:2013. Chemical disinfectants and antiseptics. Hygienic handrub. Test methods and requirements (phase 2/step 2). May 2013.

BS EN ISO 19011:2001. Guidelines for auditing management systems. November 2011.

BS EN ISO 22000:2005. Food safety management systems. Requirements for any organisation in the food chain. May 2005.

BS EN ISO 27000:2017. Information technology. Security techniques. Information security management systems. Overview and vocabulary. February 2016.

BS EN ISO 31000:2009. Risk Management. Principles and Guidelines. March 2010.

BS ISO 28000:2007. Specification for security management systems for the supply chain. December 2007.

Food & Drink – Good Manufacturing Practice: A Guide to its Responsible Management, Seventh Edition.
The Institute of Food Science & Technology Trust Fund.
© 2018 John Wiley & Sons Ltd. Published 2018 by John Wiley & Sons Ltd.

BS EN ISO 9001:2015. Quality management systems. Requirements. September 2015.

Campden BRI produce a large range of books, guidelines and research reports for every area of food manufacturing. Available at http://campden.co.uk/publications/pubs.php. These include:

Food safety plans: principles and basic system requirements. 2016. Guideline G76. ISBN 9780907503880.

Guidelines for preventing hair contamination of food – advice on head coverings. 2006. Guideline G48. ISBN 0905942779.

Guidelines for the hygienic design, construction and layout of food processing factories. 2003. Guideline G39. ISBN 0905942574.

HACCP: A practical guide, 5th edition. 2015. Guideline G42. ISBN 9780907503828.

HACCP Auditing Standard, 3rd edition. 2015. Standard. ISBN 9780907503835.

Hand hygiene: Guidelines for best practice. 2009. Guideline G62. ISBN 9780907503606.

TACCP (Threat Assessment and Critical Control Point): a practical guide. 2014. Guideline G72. ISBN 9780907503774.

The Chilled Foods Association produce a range of guidelines. Available at http://www.chilledfood.org/resources/publications. These include:

- *Best Practice Guidelines for the Production of Chilled Food*, 4th edition. The Stationery Office, London, 2006.
- *Microbiological Testing and Interpretation Guidance*, 2nd edition. 2016.
- *Microbiological Guidance for Produce Suppliers to Chilled Food Manufacturers. Micro Guidance for Growers (MGG3)*, 3rd edition. 2016.
- *Food Safety and Hygiene Training Guidance in a Multicultural Environment*, 2nd edition. 2016.
- *Water Quality Management Guidance*, 3rd edition. 2017.

Code of Practical Guidance for Packers and Importers. Weights and Measures Act 1979 Issue No. 1. Department of Trade and Industry. The Stationery Office. ISBN 9780115129223.

Codex Alimentarius: Food hygiene, basic texts. Joint FAO/WHO Codex Alimentarius Commission, Joint FAO/WHO Food Standards Programme and World Health Organization, 2003. ISBN 9251040214.

E. coli O157—Control of cross-contamination: Guidance for food business operators and enforcement authorities. Food Standards Agency, 2011.

Available at http://www.food.gov.uk/business-industry/guidancenotes/hygguid/ecoliguide#.UFsKU42PXWE.

Food Handlers: Fitness to work regulatory guidance and best practice advice for food business operators. Food Standards Agency, 2009. Available at http://www.food.gov.uk/multimedia/pdfs/publication/fitnesstoworkguide09v3.pdf.

Food Law Code of Practice (England). Food Standards Agency, 2017. Available at https://www.food.gov.uk/sites/default/files/food_law_code_of_practice_2017.pdf.

Food Safety for the 21st Century: Managing HACCP and Food Safety Throughout the Global Supply Chain, 2nd edition. Eds Wallace, C., Sperber, W. and Mortimore, S.E., Wiley, 2018. ISBN 978-1119053590.

Guidelines for the validation of food safety control measures. CAC/GL, 69. Joint FAO/WHO Codex Alimentarius Commission, 2008, pp. 1–16.

HACCP: A food industry briefing. Mortimore, S.E. and Wallace, C.A., John Wiley & Sons, 2015. ISBN 9781489986405.

Principles of Food Sanitation, 6th edition. Eds Marriott, N.G., Shilling, M.W and Gravani, R.B., Springer, 2018. ISBN 9783319671642.

Total Quality Management: The Route to Improving Performance, 2nd edition. Ed. Oakland, J.S., Butterworth-Heinemann, 1994.

APPENDIX V

CONTRIBUTION TO THE SEVENTH AND PREVIOUS EDITIONS OF THE GUIDE

GMP Working Group Members

Dr Louise Manning (Editor) and IFST Scientific Committee members

Thanks should be expressed in particular to:

- Dr Jan Mei Soon for her support with writing Chapters 5 to 7 and
- Kaarin Goodburn MBE, Director & Secretary General, Chilled Food Association for her support for the chapter on chilled foods.

This 7th edition has, of course, built on and updated the valuable content of previous editions, and substantially reflects the expertise and efforts of all those who contributed in various ways to one or more of the previous six editions.

Food & Drink – Good Manufacturing Practice: A Guide to its Responsible Management, Seventh Edition.
The Institute of Food Science & Technology Trust Fund.
© 2018 John Wiley & Sons Ltd. Published 2018 by John Wiley & Sons Ltd.